D0350938

Dodging Extinction

Dodging Extinction

Power, Food, Money, and the Future of Life on Earth

ANTHONY D. BARNOSKY

UNIVERSITY OF CALIFORNIA PRESS

University of California Press, one of the most distinguished university presses in the United States, enriches lives around the world by advancing scholarship in the humanities, social sciences, and natural sciences. Its activities are supported by the UC Press Foundation and by philanthropic contributions from individuals and institutions. For more information, visit www.ucpress.edu.

University of California Press
Oakland, California

Library of Congress Cataloging-in-Publication Data

Barnosky, Anthony D., author.
Dodging extinction : power, food, money, and the future of life on Earth / Anthony D. Barnosky.
pages cm
Includes bibliographical references and index.
ISBN 978-0-520-27437-2 (cloth : alk. paper) —
ISBN 978-0-520-95909-5 (e-book)
1. Mass extinctions. 2. Extinction (Biology).
3. Conservation of natural resources. I. Title.
QE721.2.E97B37 2014
576.8′4—dc23

2013048773

Manufactured in the United States of America

23 22 21 20 19 18 17 16 15 14
10 9 8 7 6 5 4 3 2

In keeping with a commitment to support environmentally responsible and sustainable printing practices, UC Press has printed this book on Natures Natural, a fiber that contains 30% post-consumer waste and meets the minimum requirements of ANSI/NISO Z39.48–1992 (R 1997) (*Permanence of Paper*).

For Liz, who changes so many lives for the better

Contents

Preface ix

1. The Last Ones Standing 1

2. It's Not Too Late (Yet) 17

3. A Perfect Storm 34

4. Power 50

5. Food 79

6. Money 106

7. Resuscitation 132

8. Back from the Brink 152

Acknowledgments 177

Notes 179

Index 225

Preface

I'm an optimist when it comes to day-to-day things. I tend to hope for the best and figure that where there's a problem, there's usually a solution. But I'm also a realist, which means that I think we make most of our own luck, and I'm always on the lookout for what might sneak up and bite me from behind.

I'm also a paleontologist, which means I take the long view of life as well as living day to day, and that's where I start to shift a little uncomfortably in my seat. Sometimes what seem to be insignificant events add up, until all of a sudden they are much bigger than the sum of their parts. And that's when bad things can happen really fast—as in the world changing more than we're ready for, and not for the better.

The paleontological and geological records show us pretty clearly that "bad things," at least from the perspective of keeping life as we humans like it, can take a myriad of forms. Massive volcano fields erupting and changing the chemistry of the atmosphere and oceans, asteroids slamming into Earth and barbequing most life forms that aren't insulated beneath a few feet of soil or water, major climate changes that reset the conditions for life on our planet—all of these things have happened. A very few times in Earth's history, bad things accumulated in such a way that a huge proportion of life on the planet was wiped out.

Those times are known as mass extinctions—when more than 75 percent of Earth's known species die off in a geological eye blink. To put that in perspective, imagine you were to look out the window tomorrow morning and three quarters of all the living things you take for granted were dead. Not a happy thought.

Luckily, mass extinctions are rare. They've happened only five times in the 550 million years that diverse life has occupied Earth. The reason my realist nature makes me squirm, though, is that it's beginning to look like another mass extinction—the Sixth Mass Extinction—is happening right now.

A broad body of scientific evidence says that the Sixth Mass Extinction is a very real possibility, but you don't have to be a scientist to realize what's pushed so many species to the brink today. It's us—*Homo sapiens*—and we have done it by changing the very surface of Earth, the climate, the chemistry of the oceans, and the air we breathe as we strive to support seven billion people in the manner to which we have become, or in many cases want to become, accustomed.

Those planetary changes are the by-products of many individual human achievements. Each is a justifiable source of pride for our species, but when added up, the result is that most of what we see around us is something we humans built or otherwise engineered. As a result, there are fewer and fewer places where species other than people can successfully survive on their own, a situation that will only get worse as we inevitably grow our numbers to nine or ten billion in just three short decades.

This presents a conundrum. How do we provide for the needs of people while still providing for the needs of other species? Does improving the human condition for more and more people necessarily doom other species to extinction? Or is there something we can do to avoid becoming, not to put too fine a point on it, the killers of our world—including, to some extent, of ourselves?

I've pondered those questions now for a few years—especially since I realized my kids were going to grow up and have to deal with the world we've left them. My hope is that their world will be at least as good as ours is now, and my optimistic side wants to believe that it will be.

But the realist in me says that the situation is likely to get worse instead of better unless we take a hard look at some things we take for granted that underpin the extinction crisis: power, food, and money. By power, I mean mainly how we generate the energy we need to keep the global ecosystem running at its current pace, but also, more subtly, how we exercise the power we have to make choices at the individual and collective levels. By food, I mean how to produce enough to feed billions of people for the long term, while still reserving what is needed to support other species. As for money, the key issue is how we weigh short-term gain against long-term wealth. How we deal with the essential human demands for power, food, and money will determine whether or not the Sixth Mass Extinction actually occurs. In that context, the Sixth Mass Extinction takes on a greater significance than that of saving a bunch of species for the sake of keeping the status quo. The mistakes that would ultimately kill all those other species would make life tougher for us too.

How to avoid those mistakes and their consequences for life on Earth (including human life) is what this book is about. Chapters 1 through 3 highlight how we know the extinction crisis is real. They compare past extinctions with what is happening today, set the magnitude of the current extinction crisis in realistic context, and explain why it is not too late to do something about it. Chapters 4 through 6 cut to the heart of the matter, explaining how power, food, money, and extinction are connected and how we can chart a future that breaks the fatal links in the chain. Chapter 7 is a reality check on what will be needed, in addition to charting a more productive future, to save many species that have already been driven to the brink of extinction. The last chapter explains why avoiding the Sixth Mass Extinction is so important, how we've dodged similar bullets in the past, and how we're poised to dodge this one too—if we decide we want to.

This book is a fusion of my own hands-on research on global change issues—primarily from a paleobiological perspective—and my reading and synthesis of the recent (and sometimes not-so-recent) work of many others. Those others cover a wide spectrum of disciplines, including paleontology, geology, ecology, conservation biology, molecular biology, taxonomy, food security, climatology, economics, and engineering. There are literally hundreds of individual researchers and thousands of publications that could be cited in a book like this, which weaves together information from so many different disciplines—far too many for me to include them all. My approach instead is to cite enough of the published literature to highlight that the messages in this book rest on firm scientific foundations and to give those readers interested in digging deeper an entry point for doing so. The particular studies, species, places, people, and anecdotes I've included to illustrate various points are but a tiny subset of those that could have been used. They tend to reflect my own experiences, career, and interactions with other researchers, because when all is said and done, we write best what we know best. Nevertheless, the examples I've included underscore the strong consensus of the scientific community that the Sixth Extinction crisis is real, that it's caused by us, and that it's at our doorstep.

Up for grabs in the scientific community is whether or not we can do anything to stop the Sixth Mass Extinction. After researching and writing this book, I think we can. Global society has tackled similarly big issues, sometimes succeeding spectacularly. Now, arresting extinction is both our grand challenge and within our grasp. If we succeed, the payoff will be big, not just for other species, but especially for ourselves. The optimist in me says I think we can pull it off; the realist says I hope I'm right.

Lonesome George, ten months before he died. Photo taken August 12, 2011, by the author.

CHAPTER ONE

The Last Ones Standing

Lonesome George. The name says it all. He was the last of his kind, found wandering all alone on the Galápagos island of Pinta in 1971, lumbering about in his methodical giant-tortoise way. Before then, scientists thought that his subspecies, *Chelonoidis nigra abingdonii,* was totally extinct. Giant tortoises like Lonesome George were once abundant on Pinta Island, as were other subspecies on other Galápagos islands. So abundant, in fact, that the very name Galápagos—a Spanish word describing the saddle shape of their shells—refers to them.

It was bad luck for the tortoises that first the Spaniards and pirates, and then whalers, made it a point to stop over at the islands from the 1500s through the mid-1800s. Lonesome George's compatriots became fewer and fewer as the sailors resupplied their vessels with fresh meat—tortoise meat, to be exact, the ideal food for seafarers on the go. The tortoises were big, providing hundreds of pounds of protein for each day's hunt, and even better, they were easy to catch. Hauling them on board was no simple task, though, if firsthand accounts are to be believed—"returned with five Terrapin and intirely exhoisted [*sic*]"—but once they were wrestled on deck, the tortoises were easy to store down below.[1] Put them in the hold, flip them on their backs, and stack them up. There was no need to worry about spoilage, a huge advantage in the days before refrigeration, because the animals, being tortoises, easily survived for long periods without food or water. The fact that they stayed alive meant their meat stayed fresh and tasty right up to the time they were finally killed for dinner, months after they were captured.

The need for a long-lasting food supply on those months-long or even years-long voyages made "turpining," as the whalers called tortoise hunting, a regular part of a sailor's job. The hunts were enormously effective—some sense of the number of tortoises captured comes from perusing whaling logbooks. A representative accounting comes from the logs of seventy-nine whalers that hit the Galápagos for tortoise hunting a total of 189 times in the years from 1831 to 1868: "Their combined catch during this period was 13,013.... In view of the facts that there were more than seven hundred vessels in the American whaling fleet at one time, and that the majority of these made repeated voyages to the Pacific during the above mentioned period ... it is evident that the catch here recorded was a mere fraction of the numbers of tortoises actually carried away."[2]

So no wonder Lonesome George was, well, lonesome. After he was discovered he, like those before him, was loaded on a ship, but his fate was a bit more benign. He was given a one-way ticket to Santa Cruz Island, where he was installed at the Charles Darwin Research Station, a short walk from the island's main village, Puerto Ayoro. There he calmly watched tens of thousands of tourists come and go each year, for all intents and purposes oblivious to the attractive females of another subspecies of Galápagos giant tortoise with whom the world hoped he would mate and send his genes down the line.

It was not to be. As the BBC put it in his obituary, "With no offspring and no known individuals from his subspecies left, Lonesome George became known as the rarest creature in the world."[3] When he died on June 24, 2012, most major newspapers in the world ran the story: the rarest of the rare was no more. Extinction had happened before our eyes.

I visited Lonesome George in August 2011, a few months before he breathed his last. At the time he was some years past a hundred years old—no one knows exactly how many. I was on my way back from, of all places, an oil refinery in Ecuador, where I had been helping one of my graduate students, Emily Lindsey, dig giant bones of extinct ground sloths and other long-gone species from tar-soaked sands, so I was no stranger to extinction. But when I saw Lonesome George I wasn't thinking that we were about to lose yet one more marvelous work of nature, something that evolution had taken eons of trial and error to produce. No, I was too overwhelmed by his resemblance to Yoda. Not his body—that looked like a World War II army helmet with four stubby elephantine legs sticking out of the bottom. But he had this weirdly retractable fire-hose looking thing for a neck, which was capped by a wizened, gentle face. What I realized is that all you'd have to do is stick a couple of big ears on the side of his head, morph his facial features a little bit, and there you have it: a dead ringer for the famous Star Wars sage. Now that I think about it, it would work for Dobby in Harry Potter too.

The point being, if you can imagine it, nature has probably already created it in some form or fashion, through the process of evolution. That process, and what it produced, is of course what the Galápagos Islands are best known for, beginning with Charles Darwin's accounts of his travels on the HMS *Beagle.* From September 15 well into October 1835, the *Beagle* took Darwin to Chatham (San Cristóbal), Charles (Floreana), Albemarle (Isabela), and James (Santiago) Islands. His initial impression was that it was darn hot, even inhospitable: "Nothing could be less inviting than the first appearance.... The dry and parched surface, being heated by the noon-day sun, gave to the air a close and sultry feeling, like that from a stove: we fancied even that the bushes smelt unpleasantly."[4]

But then, after he had explored each of the islands and taken note of the plants and animals that lived there and let it all settle in over the next couple of years, the lightbulb went on: Darwin realized how the species on a given island were similar to, yet not quite the same as, the species he had seen on other islands in the archipelago. He also noticed how some of the species seemed to differ from each other only in small respects. He observed this not only of the finches, which later became famous as icons of evolution, but also of the giant tortoises: "[It is possible to] distinguish the tortoises from the different islands ... they differ not only in size, but in other characters. Captain Porter has described those from Charles and from the nearest island to it, namely, Hood [Española] Island, as having their shells in front thick and turned up like a Spanish saddle, whilst the tortoises from James Island are rounder, blacker, and have a better taste when cooked."[5]

Since Lonesome George's passing, the Galápagos Islands also stand as a testament to extinction—and not just of Lonesome George. Of the reptile, amphibian, bird, and mammal species that inhabited the Galápagos when Darwin observed them, about 12 percent are now extinct, and of those remaining, nearly 40 percent are threatened with extinction. Just as Darwin saw that the Galápagos provided a scaled-down, tractable version of how evolution proceeded in the world at large, the islands now provide a mini version of how extinction is proceeding on the global scale.

And how it's proceeding is, in a word, fast. The numbers for the world stage are still a little lower than for individual island systems like the Galápagos, in part because islands tend to get hit harder and faster by extinction than continents—species on them can't leave for greener pastures when the going gets tough. Even so, the global percentages of species at risk of disappearing forever are terribly elevated for groups of animals and plants we know about, ranging from about 14 percent of bird species to about 22 percent of mammals, to perhaps 41 percent of amphibians, and to a whopping 64 percent of a type of tropical (and subtropical) plant called cycads. If we average these percentages over all of the species for which scientists have good information, about 30 percent of the world's species are threatened with extinction.

"Species threatened with extinction." Those words actually have a very precise implication to the scientists who study which species are at risk, so it is worth understanding what the words really mean when you're trying to figure out how much faith to put in the ever more frequent extinction pronouncements appearing in the news media. (Try it yourself if you haven't already: Google "extinction," then click "news" and see what you come up with.)

First, it's important to pay attention to what we mean by a species. The last time I counted, there were at least twenty-six different, in some cases mutually exclusive, conceptions of what a species actually is, and after a beer or two (sometimes even before), biologists can argue endlessly about which definition is best. But at the heart of it, most of the articles about extinction (in the scientific literature anyway) are using a pretty straightforward definition of what a species is: a group of plants or animals that can pass their genes on to their offspring, which can, in turn, pass their genes down the line to their offspring. If a group of individuals can do that, then they are grouped together as a species, or perhaps a subspecies, and that group is given a name.

How to name species can be another hot button among biologists (more beers, more arguments)—but again, for those dealing with extinction issues, naming is most commonly done in accordance with the Linnaean taxonomy you might have learned (and forgotten) in high-school biology. The species itself gets a name, and closely related species are grouped into the same genus, which also has a name. Hence, you (and I) belong to the species *sapiens,* which belongs to the genus *Homo.* Rules of nomenclature say the species name technically cannot be used without the genus name—so we are not *sapiens,* we are *Homo sapiens* (which can be abbreviated *H. sapiens*). This binomial nomenclature guards against confusion if, say, a bug taxonomist and a mammal taxonomist (who tend to run in different scientific circles and probably are not reading each other's publications) want to (or inadvertently) assign the same species name to the group of organisms they happen to be excited about at the time. That's ok as long as the genus name is different, because the two are always used together. The plural of "genus" is "genera"—genera are grouped into families, families into orders, orders into classes, classes into phyla, and phyla into kingdoms. In assessing extinction intensities, scientists often talk about the number of species going extinct within a certain genus, family, class, and so on, as I did above for species of amphibians, reptiles, birds, and mammals. Each of those major groups of animals is in fact a "class" in the formal taxonomic sense: Class Amphibia, Class Reptilia, Class Aves, and Class Mammalia.

If you do take my suggestion and Google "extinction," as often as not you will find a list of news articles about extinction not of a whole species, but of a subspecies or a population. Take this headline, from the BBC News: "Scottish Wildcat Extinct within Months, Association Says."[6] Dig deeper, and you will find that "Scottish wildcat" refers not to a species in its entirety, but to the wildcats in a part of Scotland. Wildcats as a whole belong to the species *Felis silvestris,* which is found worldwide and on balance is doing pretty well at staying alive. But there are at least five genetically distinct groups within *Felis silvestris*—and by "genetically distinct" I mean there are some very, very slight differences in their genome which cause each group to look a little different from the others. These differences arise because dispersal of animals from one region to another has been impeded for a long period of time (at least thousands of years), which means that the animals of, say, southern Europe do not often mate with those of, say, northern Africa. Through time, slight genetic distinctiveness builds up in the different geographic regions, and each of those genetically distinct groups is regarded as a subspecies. Nevertheless, if a wildcat from an African subspecies does happen to mate with a wildcat from a European subspecies, they have no trouble producing fertile offspring, which is why they are considered to remain in the same species. Subspecies can be further broken down into populations, which are groups of interbreeding individuals. The exchange of individuals between populations—which then causes exchange of genes, or gene flow, between populations—runs the gamut from very common to very rare. When gene flow is very rare, the population in question evolves into its own subspecies, and over time may even become its own species. The bottom line is that a given species is usually composed of a bunch of distinct, genetically identifiable populations, with groups of genetically similar populations lumped together and named as a subspecies.

In that case, a third identifier, the subspecies label, is added to the genus and species name—and biologists argue about what that should be, too. And that brings us back to the Scottish wildcat. Scottish wildcats, far from being their own species, are simply a group of populations within the species *Felis silvestris*. Some taxonomists regard that group of populations as so physically distinct from others that they separate them out as their own subspecies, which they named *grampia*. Thus that group of populations would formally be called *Felis silvestris grampia*. Other taxonomists regard those Scottish populations, on genetic criteria, as belonging to the widespread subspecies *silvestris*, in which case the Scottish wildcat is *Felis silvestris silvestris*.

Arcane naming issues aside, the underlying point is this: extinction of Scottish wildcats does not mean the species goes extinct. It means that a certain subset of populations that make up the species goes extinct. As sad as seeing the last of the "highland tigers" (as Scottish wildcats are called locally) would be, their loss alone would not necessarily mean the whole species is in trouble. But scale up those losses to wiping out most of the populations in most parts of the world—as is the case, for example, for real tigers—and extinction of the entire species becomes a real threat.

That is what makes the world sit up and take notice. The prelude to a species' extinction is that for whatever reasons, more individuals of that species die with each generation than are born. If that keeps up, the populations to which those individuals belong disappear—that is the stage the Scottish wildcats are at, and in fact, the disappearance of Scottish wildcats is the continuation of a trend that has already wiped out nearly all other wildcat populations in Great Britain, and most of them throughout Europe.

The next stage in the extinction process is that deaths happen too fast to be replaced by new births throughout *most* of the populations in a species. That's where the loss of Lonesome George comes in. Pinta

Island once had many populations of Galápagos tortoises in George's subspecies, but tortoise by tortoise, they faded out, until he was the last one standing. Technically, his subspecies survived until his death, but in reality, it was the walking dead, a zombie subspecies, long before Lonesome George actually died. Once a population or subspecies falls below a certain number of individuals, it can be virtually impossible to build back to viable numbers for the long run. In turn, the loss of a subspecies often reflects an overall decline in numbers of individuals throughout the entire species—as was certainly the case with the Galápagos tortoises, whose numbers began to fall precipitously after those long-ago sailors discovered them. In the 1500s, somewhere in the neighborhood of 250,000 tortoises were spread throughout the Galápagos Islands. By the 1970s, there were only 3,000 left, the result of people eating them, using and selling their fatty oils for fuel, converting their natural habitat to agriculture or other human constructs, and bringing in domestic livestock and pets that became ecological competitors or predators (particularly on tortoise eggs and baby tortoises). Even though all ten or so subspecies of Galápagos tortoises were still alive in the 1970s, the species was clearly in trouble, having declined by an average of a little over 400 tortoises a year for some 600 years. At that rate, those 3,000 individuals alive in 1970 would have dwindled to nothing in fewer than ten years if people hadn't stepped in to help.

In that context, Galápagos tortoises became (and remain) "threatened with extinction" in a strictly defined scientific sense. What scientists who study extinction mean when they use those words is much more than a general statement that a species is in trouble. It's an expression of exactly how much trouble a species (or in some cases, a subspecies) is in, defined by stringent criteria that include things like how fast individuals and populations are dying off, how fast the habitat they need to survive is disappearing, and so on. The scientific body that has taken on the task of determining extinction risk is composed of thousands of biologists and is known as the International Union for the Conservation of Nature, or IUCN. Without going into detail here, suffice it to say that the IUCN has come up with a set of rules for sorting species into categories according to their risk of extinction.[7] Species are listed by category in the IUCN Red List of Threatened Species. Those of "least concern" are considered to be in pretty good shape. Subsequent categories move to progressively higher risk of full-on extinction: "near threatened," "vulnerable," "endangered," "critically endangered," "extinct in the wild" (meaning the only survivors are in zoos and such), and finally, when the last individual blinks out, "extinct." Species that fall in the three categories most at risk, short of being extinct in the wild—vulnerable, endangered, and critically endangered—are grouped under the IUCN label "threatened."

This raises an important flag when you're reading about extinction. Like many scientists who write about these things, when I call a species "threatened," particularly in this book, I'm using the word in its IUCN context—it's shorthand for saying that a species falls in one of those three higher-to-highest risk categories. That, of course, is not always the case when you read a news article about extinction, in which "threatened" may be used in a much more general sense. Likewise, you have to be careful with the word "endangered," because it too is used in many different ways. For instance, saying a species is endangered under the United States Endangered Species Act is not quite the same as saying it's endangered in the sense of IUCN categories.

The IUCN has its work cut out for it. Studying all the species in the world to determine which of the categories they fall into is a Herculean task—by latest estimates, there are some 8.7 million species in the world (and that doesn't count the vast majority of the microscopic ones). As of 2013, the IUCN had worked its way through about 70,000 species, at the rate of about 5,000 species each year over the preceding decade. It's not too hard to do the math and realize that even if we were to assess all of the species that scientists have named—which amounts to only about 1.3 million out of those 8.7 million estimated to exist—we'd be looking at another 246 years of work. So the IUCN has come up with two other categories to sort species into: "not evaluated" and "data deficient."

Most species are in the "not evaluated" category—simply put, nobody has taken the time to look at them yet, so we don't know if they are in good shape, bad shape, or somewhere in between. "Data deficient" species are those for which we have some inkling of what's going on, but which haven't been studied enough for us to stick our necks out and place them in one of the other categories. Which brings up an important point about the way the scientists who categorize species in this way work: they tend to be cautious about saying something is so if there is not strong evidence to back up the conclusion.

The problem with that methodical approach to sorting species into categories is that it sometimes puts a not-so-bad-sounding label on an on-the-ground situation that is really pretty terrible. Take "vulnerable," for instance, the category into which Lonesome George's species (*Chelonoidis nigra*) is placed. What doesn't really come through with that simple word is that by the 1970s, the number of animals that used to be normal for the species had been reduced by 99 percent.

African elephants (*Loxodonta africana*) are also classified as "vulnerable" by the IUCN as of 2010 (the date of their most recent comprehensive evaluation). That listing takes into account that ivory poachers killed nearly half of Africa's elephants in the 1980s, before an international ban on ivory trade quelled the market. It does not take into account what has happened since 2009: as affluence has increased in China, the demand for ivory there has driven the price sky high, making ivory sales profitable enough for international crime rings to get into the act. As a result, the 50 percent of African elephants that managed to survive the 1980s onslaught are being decimated in record-high numbers. In Tanzania alone, 31,348 elephants fell under poachers' guns from 2009 to early 2012—that's 42 percent of Tanzania's entire herd.[8] In the space of just a few weeks in early 2012, marauding horsemen killed and hacked off the tusks of over half of the 400 elephants in Cameroon's Bouba N'Djida National Park.[9] In Garamba National Park in the Democratic Republic of the Congo, 22 elephants—a whole extended family—were slaughtered in one fell swoop by gunfire that appears to have come from a Ugandan army helicopter. By the time rangers got there, a million dollars' worth of tusks had been hacked off and hauled away.[10] That event, on top of the kills that preceded it, contributed to reducing the total number of elephants in Garamba to a tenth of the previous population, from around 20,000 to 2,400. The new twist is dumping cyanide into water holes at which elephants drink, a technique that killed 81 elephants in a single September day in 2013 in Hwange National Park, Zimbabwe.[11] The net effect is the greatest percentage loss of elephants in history. More than a million elephants are thought to have roamed Africa prior to 1980; after the 1980s decimation there were perhaps 600,000 left; in the three years preceding 2012, that dropped down to about 450,000. Continuing the death rate that has marked the years 2009–2012 would mean there would be no more wild African elephants on Earth in twenty years. That's "vulnerable" to extinction indeed.

Now, multiply that elephant story, or Lonesome George's story, by 10,519—that's how many species the IUCN has classified as vulnerable as of the time I write this.[12] Even worse off, in terms of extinction risk, are those species in the endangered category—add another 6,169 species. Endangered species are those like the Atlantic bluefin tuna (*Thunnus thunnus*), which we love to eat and whose numbers and geographic extent have been reduced more than 50 percent in their last three generations (that is, over about three decades); the blue-capped hummingbird (*Eupherusa cyanophrys*), a stunningly beautiful iridescent green bird with bright purple wings that now is down to about 2,500 individuals restricted to a small area in southern Oaxaca, Mexico; and the growling grass frog (*Litoria raniformis*), whose numbers have fallen by more than half in the last decade, thanks mostly to a deadly fungus that is killing amphibians by the millions worldwide. Critically endangered species—numbering 4,226 as I write this—are even closer to the brink. Examples include western gorillas (*Gorilla gorilla*) and the bizarre-nosed chameleon (*Calumma hafahafa*) (a species that deserves keeping alive for its name alone). Gorillas have lost 80 percent of their individuals in just the last three generations (over the past sixty years); even in protected areas, 45 percent of them died in the twenty-year period from 1992 to 2011. Bizarre-nosed chameleons are known only from Madagascar, where a very few individuals have been found exclusively in small patches of habitat that probably total no more than a hundred square kilometers. That habitat is rapidly being taken over by slash-and-burn agriculture, logging, and cattle grazing.

Rapid losses like these are rampant. The number of species we now know to be threatened with extinction (remember, that's the sum of all the vulnerable, endangered, and critically endangered species) is a whopping 20,614 out of the approximately 70,000 species that have been assessed by IUCN standards. That list includes not only animals with backbones (that is, vertebrate animals), which I chose for all of my examples above, but also annelids (worms and relatives), arthropods (insects, spiders, and relatives), cnidarians (jellyfish and relatives), molluscs (clams and snails), nemertine marine worms, onycophorans (velvet worms), plants, and a very few protists.[13] All of these species are now on the Lonesome George trajectory—their remaining individuals are the last ones standing after intense human impacts have severely depleted their numbers over the past few decades.

Let's say, for the sake of argument, that all of these species continued down that Lonesome George path, eventually dimming into extinction. Just how abnormal would that be, given that Earth's species have waxed and waned over much longer courses of time than just a human lifetime or two? Here's a way to get a quick sense of how what's happening now compares to what's happened over the past five centuries or so. We know that about eight hundred species have died out in the last five hundred years. We also know that the vast majority of those species were driven to extinction by people—because we overhunted them, converted their habitat into something that we thought would serve us better in the short term, or introduced competing species intentionally or inadvertently. Those are, of course, the same threats that are driving species toward extinction today, although the magnitude of those threats has increased dramatically, and some new ones (like pollution and climate change) have been added. Continuing the present rates of decline means that all of those 20,614 currently threatened species will be extinct within the next five hundred years (many of them much, much sooner). That would be twenty-five times as much loss in the next five centuries as in the last five.

Considering that even the past five hundred years—the years 1500–2000—had elevated extinction rates with respect to what's normal over the thousands and millions of years that make up geologic time (as I'll deal with in more detail in chapter 2), what appears to be coming down the pike in the next few centuries is truly out of the ordinary. So out of the ordinary, in fact, that sustaining current rates of extinction will guarantee that we bring on the so-called Sixth Mass Extinction. It's called that because there have been only five other times in the past 540 million years (that's right, million years) when extinction on Earth has been as intense and as rapid as it is today. Those are known among paleontologists as the Big Five Mass Extinctions. The most recent was about 66 million years ago, when the big dinosaurs were lost forever, and that was in fact one of the least severe. "Only" 76 percent of Earth's known species died out then. The worst mass extinction was about 252 million years ago, and it saw the deaths of more than 90 percent of known species. No wonder it's been called the Great Dying.

That's why the thought of the Sixth Mass Extinction—where we're headed if we do nothing to save all the species presently at risk—is such a big deal. I'll elaborate on our progress toward the Sixth Mass Extinction in the next chapter, but for now, it's enough to say we really don't want to see it occur. We'd be looking at a world where more than three out of every four of the species we take for granted would be gone.

That doesn't have to be. There are rays of hope. The last individuals standing as threatened species today don't have to be dead species walking, despite their high-risk status, because what the IUCN categories really reflect is not inevitable extinction, but "the likelihood of a species going extinct *under prevailing circumstances*" (italics mine).[14]

Those prevailing circumstances are the tremendous pressures humans put on other species through our sheer numbers and attendant needs and wants, which, as past experience shows and as business-as-usual would continue, leads to habitat destruction, overhunting and overfishing, introduction of competitors or predators that kill native species, pollution, climate change, and, importantly, the interactions among all those things, which act as multipliers of those threats. Predictions of doom all have one key assumption: that we will simply keep on doing things as we always have, without any regard for what we now know. That's kind of like being a train operator and seeing a school bus stalled on the tracks way off in the distance, knowing you can stop in time if you pull the brake lever hard now, but deciding what the heck, let's not bother. That would be an unusual scenario—humans are, in most cases, smarter than that.

And in fact, we've shown we can be smarter than that when it comes to preventing extinction. Which returns us to Lonesome George and his species. Recall that in the 1970s, when Lonesome George was found, there were only 3,000 members of his species left, the nadir following the 99 percent loss that had begun centuries before. Now, more than 19,000 tortoises are doing their thing in the Galápagos. What reversed the almost-guaranteed extinction trajectory?

In a word, people. Humanity saw the train wreck coming and decided to do something about it. The first step, of course, is acknowledging that the wreck is imminent. In the case of Galápagos tortoises, that was already evident by 1936, when the government of Ecuador listed the tortoises as a protected species. That acknowledged the problem, but tortoise numbers continued to dwindle, in part because the pressures on the tortoises had shifted from buccaneers eating them to settlers converting their habitat to farmland; there was also a decent market for selling the tortoises, their eggs, and their shells to dealers outside Ecuador.

By 1959, Ecuador had addressed those problems in part by declaring all places uninhabited by people in the Galápagos a national park, and the Charles Darwin Foundation was created under the auspices of UNESCO and the World Conservation Union with the mission of preserving Galápagos ecosystems. The net effect was to elevate the national park to a global treasure and turn it into an economic engine for the locals and for Ecuador in general. Protection for tortoises and other species was strengthened at the same time: it became illegal to capture or remove any species from the islands. Nevertheless, the profit in tortoise trade did not go away, and tortoises and their eggs continued to be sold clandestinely, prompting Ecuador to pass another law in 1970 that made it illegal to export Galápagos tortoises or their eggs, regardless of whether the specimens were captured in the wild or bred in captivity. Finally, in 1975, the market was largely cut off when the tortoises were listed as an endangered species, not by the IUCN, but by the Convention on International Trade in Endangered Species of Wild Fauna and Flora, commonly abbreviated as CITES.[15] That convention, by international agreement, places strict prohibitions on the import and trade of listed species—including by-products such as shell or bone—for all of the ratifying states, which now include about 179 out of the 196 nations in the world.

All those legal measures slowed the killing of the last tortoises on the islands but couldn't change the fact that 99 percent of the species was gone. There was one more piece to the puzzle of how to save them, and that was to make sure more of their young survived under the conditions in which the tortoises now found themselves: a world where introduced animals like feral goats, dogs, and pigs ravenously ate tortoise eggs as well as baby tortoises. The way around that was for people to step in, collect the eggs prior to hatching, and take them to captive breeding colonies to rear the young past the age at which they were easy prey—it doesn't take too many years for the tortoise shells to get big enough and tough enough to provide the right kind of protection. Once that happened, the tortoises were (and are) returned to where the eggs were laid, and off they go on their own in their natural habitat. In concert with building the captive breeding program, the Galápagos National Park Service began to find and kill the introduced predators.

As a result, the tortoises are coming back. The day I saw Lonesome George at the Charles Darwin Research Center in 2011, I also saw, in enclosures not too far from him, young tortoises of most of the Galápagos subspecies, sorted by age and island, that were being groomed as saviors of the species. That this process is working became pretty clear to me when I walked along a windy beach on Santa Cruz Island the next day, waves crashing, and found myself dodging tortoise nests, some of which retained the leathery remnants of hatched eggs at their edges. The increasing numbers mean that Lonesome George's species is scrambling back from near annihilation. The tortoises are not out of the woods yet, but they have climbed one IUCN category in the right direction, from being endangered in 1996 to being vulnerable in 2012.

That kind of success makes it clear that just as we humans can put tens of thousands of species at risk of extinction, we also have the power to pull species back from the brink, using the right combination of human ingenuity, carrots, and sticks. Multiplying that one success story by 20,000 or so more species is a pretty tall order. But, as we'll see in the rest of this book, it is within humanity's grasp, and we have good reason to make all those new success stories a reality—they're all that stand between us and the Sixth Mass Extinction.

CHAPTER TWO

It's Not Too Late (Yet)

On July 19, 1978, our four-wheel-drive, three-quarter-ton pickup truck rolled into Edgemont, South Dakota, just about sundown. It fit right in. With less than 1,000 people, Edgemont is not unlike many small rural communities in that part of the northern Great Plains, a cluster of simple frame houses and a quiet Main Street that grew up next to the railroad yards, waxing and waning through the years as wool mills, munitions factories, and uranium mines came and went. It is, however, absolutely remarkable in one respect: the number of mosasaurs per square mile in the surrounding ranchlands is probably greater than the number of people. Mosasaurs were huge marine lizards with fins instead of legs. They have been extinct now for some sixty-six million years, but their fossil bones litter the countryside around Edgemont.

Those bones were what had caused us to beeline out there, a twenty-six-hour drive from Seattle. At the time, we were a motley crew of budding paleontologists who were happy to escape the confines of the basement of the University of Washington's Burke Museum to do some fieldwork. The pickup truck belonged to Jim Martin, who, like the truck, was big—easily six-five with shoulders to match, a big white cowboy hat, a fist-sized belt buckle, and in those days, a Fu Manchu mustache. Jim is a native South Dakotan, and it was his knowledge of the local geology and landowners that brought us to Edgemont. His love of that country took him back to Rapid City to make major paleontological contributions at the South Dakota School of Mines and Technology over the next thirty years.[1]

Back in those days, Jim and I were office mates and classmates, both working on our graduate degrees in geological sciences and paleontology. He was already a veteran fossil hunter, but the trip to Edgemont was the first hard-core paleontology expedition I had ever been on, and I was excited to be going after big game, which mosasaurs most definitely were. Although not dinosaurs in the technical sense, mosasaurs were dinosaur sized. The largest were nearly sixty feet long, which means they'd hang over the edge of the eighteen-wheel semi trucks you see on the interstate, and they had lots of huge sharp teeth, features that have earned them the nickname "*Tyrannosaurus rex* of the sea." In fact, their closest living relatives seem to be snakes (in contrast to the real *T. rex,* whose closest living relatives are birds), and instead of terrorizing their prey on land, mosasaurs swam and hunted in long-ago oceans.

What made mosasaur skeletons so common around Edgemont was a geological convergence that eventually became the Pierre Shale (pronounced "peer" by those in the know), the solidified, tectonically tortured remains of gray, almost black mud that accumulated on the floor of a vast shallow sea.[2] That so-called Western Interior Seaway once covered not only the part of the world where Edgemont now stands but also the whole eastern Rocky Mountain front and Great Plains, and it connected the Gulf of Mexico with the Arctic Ocean. Mosasaurs were abundant in those seas, and when they died, their remains bloated and floated around for a bit, then sank to the bottom. The flesh decayed, but the bones became entombed for all time in the oxygen-poor mud. Sediment eroding from the uplands to the west then washed into the sea and over the course of tens of thousands of years buried the mosasaur graves so deep that the enormous pressure and heat were enough to turn gray mud into gray rock and buried bone into white stone. The ancient, lithified muds and the now-fossilized mosasaur skeletons they preserved were millions of years later thrust up to the surface as the Rocky Mountains began their relentless rise, causing the seas to retreat and the Great Plains to emerge.

The midsummer landscape around Edgemont today is an artist's study in shades of gray and green, with the gray being the Pierre Shale substrate and the green the vegetation that manages to conquer it in a few places. As paleontologists, we headed right for the gray gullies—that's where we expected the bones to be eroding out. And we weren't disappointed. Within a couple of days, by keeping our eyes peeled for white-colored bits of broken fossil bone flecking that monotonous gray shale, we had found skeletons of a least two gigantic mosasaurs, the wing bones of a pterosaur (a big flying reptile) that at one time swooped above the waters where the marine lizards swam, and many other scraps of oceangoing critters that lived and died at the same time as the dinosaurs. It wasn't as easy as it sounds, because once you locate those white flecks, the next part of the job is akin to working on a chain gang—wielding a sharp-pointed pick against solid rock in the hot sun, over and over; cleaning the surface off with a whisk broom or paintbrush; smacking the rock again, chipping away the shale that encases the fossils, piece by hard-won piece. It doesn't help that the Murphy's Law of finding skeletons is that they always lie such that they are heading directly into the heart of the hill, instead of parallel to the eroding hill slope.

You can't go through an experience like that without wondering why there are no mosasaurs alive today. At least I couldn't. Luckily, the answers were to play out before my eyes in a scientific controversy that would continue over the next three decades and that still raises hackles in some circles today. I already knew that mosasaurs were one of the casualties of the most recent of the Big Five Mass Extinctions, part of the 76 percent of Earth's known species that died out sixty-six million years ago. What I didn't know, nor did anyone else at the time, was what actually caused such an enormous dying off. A discovery that would go a long way toward explaining that is what generated all the controversy.

That last huge extinction event took place at the end of the Cretaceous period. The Cretaceous is a segment of geological time that lasted from about 145 million to 66 million years ago and, like other geological periods, is clearly recognizable by particular groups of species found as fossils. In fact, the geological periods were originally defined when geologists of the 1700s and 1800s noticed that the lowest rocks they banged their hammers against contained a certain assemblage of fossil species, the next layer up a different assemblage, and so on, and that no matter where they were in western Europe, they saw the same progression of fossil assemblages as they climbed from the lowest to the highest layers in any given stack of rocks. Thus they saw how life evolved through time, for the oldest rocks are almost always on the bottom.[3] For most of the geological periods, the early geologists concentrated on marine invertebrate fossils, for the simple reason that marine rocks—that is, those formed like the Pierre Shale, from sedimentary particles that accumulated on the ocean floor—were what happened to be in the areas in which they were working. The Cretaceous was no exception: the distinctive fossils included extinct species of clams, snails, ammonites (relatives of the living chambered nautilus), and tiny calcium carbonate shell parts of microscopic algae (called coccolithophores) that form the White Cliffs of Dover and are found in matching rock formations on the other side of the English Channel. In some of those rocks, mosasaur bones were right alongside the fossils of clams, snails, and other ocean-dwelling creatures that lack backbones.

Similar marine critters turned up on the other side of the Atlantic when people started looking there—including in the Pierre Shale. This is how geologists knew that the Pierre Shale not only was Cretaceous in age but could be slotted into the last quarter of the Cretaceous. Once methods to date the minerals in rocks by measuring their radioisotopes were developed, it became possible to estimate how many years old the Pierre Shale around Edgemont, and the mosasaur bones encased in it, actually were.[4] It turned out those mosasaurs from the Pierre Shale lived and died between 72 and 84 million years ago.

At the same time that the Pierre Shale mosasaurs were swimming around in the ocean, on land it was the Age of Dinosaurs, which at that point had been going on for about 170 million years. The dinosaurs were stomping along a belt of higher ground to the west that ran from Canada down to New Mexico. They included the big boys like *Pentaceratops* (the sumo wrestler type with the big neck frill), *Maiasaura* (the "good mother" dinosaur that famed paleontologist Jack Horner popularized), *Alamosaurus* (the largest known dinosaur from North America, looking like a long-necked, long-tailed brontosaurus), and the fearsome predator *Albertosaurus* (a *T. rex* cousin). The descendants of these, as well as of the mosasaurs we found in the Pierre Shale, continued to thrive for the next seven million years or so, right up until the end of the Cretaceous, when the mass extinction event wiped them out. After that, an entirely different kind of animal—mammals—became dominant on land, and the Cretaceous gave way to a new geological period, the Paleogene. Because geologists abbreviate the Cretaceous as K and the ensuing Paleogene as Pg, the mass extinction event that took place near the boundary between the two geological periods is known as the K-Pg event.[5]

It was study of the chemical signature left at the K-Pg boundary that gave rise to all the controversy. About the time I was collecting mosasaurs in South Dakota, a UC Berkeley geologist named Walter Alvarez was collecting foraminifera—tiny fossils of microscopic marine protozoans—from rock outcroppings in a roadcut near Gubbio, Italy. The roadcut was long known as a place where the K-Pg boundary was well exposed, and Walter knew that as one result of the mass extinction event, certain species of the foraminifera suddenly disappeared from the section, leaving the thin clay layer right at the boundary devoid of the little fossil protozoans. Right above the boundary clay layer, voilà, relatives of the extinct species began to show up again. What Walter didn't know, and what he wanted to find out, was how long it took for those new species to show up. The standard geological dating methods of the time simply weren't up to the task—they couldn't (and still can't) easily resolve how long it took just a few inches of sediment to accumulate.

Many years later, after I joined the faculty at UC Berkeley, I met Walter, a sandy-haired guy with a ready smile and blue eyes that sparkle with inquisitiveness. What immediately became clear is that he is never short on big ideas, loves to discuss them, and is really, *really* good at thinking outside the box. Which led him, I suppose, to talk over his dilemma with his dad, Nobel Prize winner Luis Alvarez, who worked up the hill from the university as a physicist at Lawrence Berkeley Laboratories. They decided that one way to ascertain how long it took for the clay layer to be deposited—and thus how long it took the foraminifera to recover after the extinctions—might be to measure the concentration of a rare element called iridium in Walter's rock samples. They would see if there was any variation in the concentrations below the K-Pg boundary, through the boundary clay itself, and above the boundary. This, they hoped, would be helpful in estimating elapsed time in the rock record, because the main way iridium gets to Earth is by raining down, in minuscule amounts, as comet and asteroid dust, presumably at more or less constant rates. By looking at variations in iridium concentrations across the K-Pg boundary, they reasoned, they might be able to roughly calculate how many years the boundary clay spanned, which would in turn tell them how long it took those foraminiferans to repopulate the ancient sea after the extinction event.

Like most great ideas, it didn't pan out. But it did lead to an even more interesting discovery: they found way more iridium in that boundary layer than there should have been. The most reasonable explanation was that an iridium-laden asteroid had crashed into Earth, causing a huge explosion that distributed iridium-rich dust around the world. The dust then settled and forever marked that instant in the geological record. It seemed likely to them that an explosion that big would be capable of wiping out much of life on Earth, and that's what the Alvarez father-son team and their colleagues Frank Asaro and Helen V. Michel proposed in a landmark paper entitled "Extraterrestrial Cause for the Cretaceous-Tertiary Extinction," published in the journal *Science* on June 6, 1980.[6]

When I read that, sitting at my graduate student desk at the University of Washington, my immediate reaction was, "No way. They're crazy." And I wasn't alone in thinking that—paleontologists can be a pretty conservative bunch, trained as we were in those days to bow down to the Law of Uniformitarianism, first conceived by Scottish geologist James Hutton in the late eighteenth century and popularized by the father of geology, Charles Lyell, in his 1830 treatise *The Principles of Geology*. In its most basic form, the law says that everything we see on Earth today can be explained by the gradual accumulation, over eons of geological time, of the sorts of tiny changes we observe going on around us.[7] Take, for example, the Grand Canyon. The Colorado River would have to have eroded its channel only four one-hundredths of an inch per year over the seventeen million years that many geologists think have elapsed since the river started to run through that part of Arizona. That is a tiny yearly change indeed, but with big consequences—a canyon more than a mile deep, and one of the world's natural wonders.

In that context, much of the vertebrate paleontology community in the 1980s regarded the asteroid impact theory as the stuff of science fiction. I didn't have a dog in the fight—by that time I had focused my graduate research on fossil mammals, studying, of all things, how extinct moles dug (why that was of interest is a long story)—but those Cretaceous mosasaurs I had excavated a couple of years prior had stayed at the back of my mind. Could this asteroid idea really explain why animals that had ruled Earth for so many millions of years could disappear in what seemed to be a geological eye blink? Of course, the Pierre Shale mosasaurs I was remembering had lived and died a few million years before the extinction event, but nevertheless, their descendants, different species filling the same ecological niche and looking pretty much the same to the casual eye, persisted right up until whatever happened at the end of the Cretaceous changed the world forever.

So I followed the arguments pretty closely. It wasn't just resistance to the idea that major extinctions could be caused by something that came from outer space that troubled many paleontologists. The main problem was that what we knew about dinosaurs just didn't seem to fit. Back in the 1970s and 1980s, the most detailed information about dinosaur life and death was coming out of a place called Hell Creek in Montana. Coincidentally, it was another Berkeley professor, whose office and lab were in the same building as Walter Alvarez's, who was accumulating that data. William A. Clemens (or Bill, as he likes to be called) had been taking crews into the Hell Creek area for a couple of decades by the time the asteroid impact article hit the newsstands. I first met Bill around that time, at a paleontology meeting held in Berkeley. He struck me as the quintessential Berkeley professor, distinguished looking with his gray hair and well-trimmed beard, speaking thoughtfully with a distinctively sonorous voice. We later ended up as colleagues in the same department, and I came to appreciate even more his careful, measured approach to science and his graciousness in putting forth and listening to ideas.

Bill's research focused not on the dinosaurs per se but on the seemingly insignificant little mammals that lived alongside (or more accurately, underneath) the big beasts. He was trying to figure out how the stage was set for mammals—culminating with us—to become the dominant life forms on Earth after the dinosaurs died out. That task required him and his graduate students to painstakingly map the sequence of late Cretaceous rocks named the Hell Creek Formation (does it give you a mental picture of the place?), measuring how far each dinosaur and mammal fossil they found was from the K-Pg boundary, which as luck would have it was exposed in that part of eastern Montana. The data they had accumulated over the years just didn't seem to fit with the explanation that the dinosaurs died out over a bad weekend, because instead of seeing a highly diverse dinosaur fauna right up to the K-Pg boundary, the fossils seemed to tell a story of gradual dwindling over about ten million years, so that there were relatively few species left to go extinct at the end. Put another way, instead of the dinosaur family tree going up all at once in a gigantic fireball, it looked like the tree had been gradually pruned over a very long time, until only a few twigs were left to die of who knows what at the end.

Bill and his recently graduated PhD student J. David Archibald, who by then had taken a job at Yale University, responded to Alvarez and company by publishing an article in 1982 that explained how the dwindling of dinosaurs they saw in the Hell Creek area was at odds with death by asteroid. They also pointed out other inconsistencies, such as the mismatch in timing between plant, land animal, and marine animal extinctions, and why some species you would expect to croak if an asteroid had hit had made it through the extinction event just fine.[8] Another objection was that at the time, we just weren't certain enough about how iridium became concentrated in rocks to know if it was foolproof evidence of an asteroid impact. And, finally, many paleontologists were asking, where was the crater? An asteroid strike that big certainly should have left a huge hole in the ground, evidence of which should have withstood the ages.

Meanwhile, there had been a rush to measure iridium concentrations at other places where the K-Pg boundary was exposed, both in marine rocks (those deposited on the ocean floors) and terrestrial rocks (those deposited on land). The spike in iridium at the K-Pg boundary kept rearing its ugly head—it wasn't just a one-off occurrence at Gubbio, Italy, and it was hard to explain without an extraterrestrial event. Some paleontologists fired back at the ideas Bill Clemens and Dave Archibald had laid out, suggesting that when the Hell Creek dinosaur record was corrected statistically to account for the vagaries of the fossil record, the purported ten-million-year (or so) gradual decline of the dinosaurs didn't hold up.[9] Bill and Dave pointed out potential problems with those alternative interpretations, and on it went.

And then, in 1991, the smoking gun was found by a graduate student, Alan R. Hildebrand, and his colleagues from the University of Arizona, Harvard, and Pemex, the Mexican oil exploration company.[10] They discovered a crater on Mexico's Yucatán Peninsula. Dubbed the Chicxulub Crater, it was big enough, seemed to be the right age, and had escaped notice up until then because half of it was underwater in the Gulf of Mexico and the other half was hidden by jungle vegetation. In retrospect, the crater is obvious if you look down from the sky—the land-based half is demarcated by a half-ring of cenotes (collapsed limestone waterholes) that arc through the northwestern half of the Yucatán, and the half submerged in the Gulf shows up in seismic and core profiles the oil companies assembled. By this time, I was serving on various faculty committees with both Walter and Bill, and when the committee business got boring (as was often the case), I got to see scientific discourse at its best—respectful, good-humored interchanges, with Walter pushing a big impact, Bill throwing paleontological roadblocks in his way, and each trying to reconcile what clearly was strong, valid evidence from the other side.

There was another sticking point, though. Although by the early 1990s it was getting harder and harder to discount the probability that a big asteroid had indeed crashed into Earth around the time the last dinosaurs disappeared, there remained the problem of why many different species were able to survive such a cataclysm. Birds made it; so did turtles. So did crocodiles, not so hugely different from their dinosaur cousins, and many kinds of plants, insects, and of course, mammals. That brought yet another paleontologist into the fray: the late, great Malcolm C. McKenna. Malcolm was the closest thing to a renaissance man I knew. As he did with many students, he stepped in at critical times to inspire me with his breadth of knowledge, his willingness to lend a helping hand, and his seamless transitions between tromping through the high mountains of Montana, a week of stubble on his face, to locate a key fossil site, to sipping gin and tonics at sundown, wearing a jacket and tie on the deck of a cruise ship in Dublin Harbor.

Malcolm got into the K-Pg act because he wasn't satisfied with the vague explanations of how an impact would play out to mass extinction. At the time, most of the ideas thrown out involved some combination of blocking out the sun with dust raised by the explosion, which would cause something akin to nuclear winter, and just blowing everything to hell. Malcolm knew a lot about the Cretaceous; he was a graduate school classmate of Bill Clemens's, and later he (with his wife, Priscilla, his partner in crime on many research campaigns) led expeditions to collect dinosaurs and contemporaneous mammals from famous localities in Mongolia and many other places. He put together a team that included a theoretician, an atmospheric chemist, and other paleontologists and geologists who had worked on Cretaceous problems, and tried to figure out what would have really happened if an asteroid of the size required for carving out the Chicxulub Crater had crashed into Earth, and moreover, why the survivors survived.

The result was an article, published in the *Geological Society of America Bulletin* in 2004, that soon became known among paleontologists as the "Cretaceous Barbeque" paper.[11] It was nicknamed that because, through credible calculations and modeling efforts, the team's work demonstrated that one of the main effects of a Chicxulub-style asteroid impact would be to release so much energy that most of Earth's surface would be superheated, such that anything that wasn't insulated by at least a couple of feet of soil or water, or otherwise protected, would be cooked in short order. This mechanism explained why many species made it through the impact—survivors tended to be animals with lifestyles that typically kept them underground or underwater, or plants whose seed banks or roots were deep enough to withstand the superheating.

That sounded plausible, but I kept thinking of those mosasaurs I had excavated from the Pierre Shale. After all, a good number of them must have been swimming deep enough in the sea to be insulated from the blast and resulting effects on surface waters. To explain their extinction at the K-Pg boundary, one more puzzle piece had to be fit into place. Ecologists have long known that marine plankton—tiny animals and plants that drift in the current and, when they are really abundant, turn the water soupy green—form the base of the food chain. Those plankton live primarily in the photic zone, that is, the top of the water column where light, not to mention the heat and debris from an asteroid blast, can penetrate. That means that when the K-Pg asteroid hit, most of those plankton were immediately poached, and those that survived had to do so in a world where life's ground rules had been dramatically reset: a new normal for the global carbon cycle; major changes in ocean and atmospheric chemistry; rapid climatic changes for years to decades to come, as particulates in the air partially blocked the sun's rays and climate-forcing gases released by the blast changed atmospheric circulation; and over the shorter term, things like muddied surface waters from all the sediment eroding from the charred and blasted land. The sudden loss of so much plankton would mean that everything higher in the food chain, mosasaurs included, was suddenly also in deep trouble. In effect, the asteroid killed in two ways: directly, by superheating the planet, with attendant catastrophes like wildfires, dust-choked skies, and so on; and indirectly, by cutting links out of the food chain, causing major climate changes, and perturbing the global geochemical cycles on which life depended.

It took several more years of back-and-forth to establish some key details, but finally the jury was in, as summarized in a report published in *Science* in 2010.[12] In a letter published in *Science* a few months later, even those paleontologists who were skeptical that an asteroid told the whole story acknowledged that "we have little doubt that an impact played some role" in the K-Pg extinction.[13] Even more evidence that the asteroid contributed heavily to dinosaur demise came from a study led by Berkeley geochronologist Paul Renne. Renne's study nailed the asteroid impact to within 32,000 years of the mass extinction—an amazingly precise match given the difficulties of geological dating of events that old.[14] Controversy still remains about whether the asteroid took out a dinosaur fauna much reduced from its heyday by a prolonged stretch of other bad luck, which is the explanation preferred by many paleontologists who study late-Cretaceous dinosaurs and mammals. But one thing seems pretty certain: the asteroid did indeed hit, and it had lasting impact. The stage was set for new life forms to evolve and dominate the world, and over the ensuing sixty-six million years, we—*Homo sapiens*—eventually did.

We did so to the extent that today, *we* are the asteroid. The human race is impacting the planet in ways that are every bit as dramatic and lasting as what happened when that big rock fell out of the sky at the end of the Cretaceous. One big difference, though, is that, unlike the dinosaurs, we can see what's coming. The other big difference, of course, is that this time it's not something that is being done to us; it is something we are doing to ourselves. As we'll see throughout this book, our impacts on the planet come from three main directions. The first is how we power our existence, by which I primarily mean how we get our energy and how we transfer energy through the global ecosystem, although another kind of power also has impact—the power we have to exercise choices and take actions. The second impact is where we get our food. And the third is how we make our money. Ultimately, it's how each one of us acquires those necessary components of human survival, multiplied by the seven billion of us now on the planet, plus two or three billion more who will be added within the next thirty years, that becomes the asteroid-sized problem for life on Earth.

Equating human impacts with an asteroid impact may at first seem far-fetched. It did to me when I first started thinking about the 1,000-plus species we've already driven to extinction and the 20,000-plus species that the IUCN regards as threatened (a number that continues to grow), and how those might compare with extinctions over the vastness of geological time.[15] Sure, the examples and statistics I mentioned in the previous chapter are gut-wrenching, but the paleontologist in me had a sneaking suspicion that if we were just able to take the same kinds of information we use to calculate extinction rates and magnitudes over the last few human lifetimes and compare it fairly to the fossil record using similar taxa and techniques, the current extinction crisis wouldn't measure up to the past Big Five extinctions. It seemed like most of the comparisons between present and past that had been made by scientists were like comparisons of apples and oranges—different kinds of species were compared, from different parts of the world and different ecological settings, and using different ways to express extinction rates and magnitudes.

In hopes of figuring out whether or not we truly were seeing the Sixth Mass Extinction bearing down on us like that Cretaceous asteroid, in 2010 a group of us in the Integrative Biology Department at UC Berkeley began to compile the information we needed to make an apples-to-apples comparison. We started with the IUCN data, then mined paleontolological databases and literature to determine which kinds of species were well enough known in both the fossil and modern record to reasonably compare. That narrowed the field to mammals very fast. The IUCN has assessed all of the known living mammal species—which is not the case for most other kinds of organisms—and the fossil record is rich and well described for mammals because bones and teeth, being hard and resistant to decay, preserve well once they are buried.

We then developed ways to calculate extinction rates that accounted for the fact that computing a rate over a short time—say, a few centuries—can give you a misleadingly high rate in comparison to one measured over a very long interval, like a few million years. Once we understood how to deal with that, we were able to determine the maximum rate of background extinction for mammals over the past sixty-six million years—that is, the normal balance between origination of new species and the extinction of existing ones. We were careful to compute that background rate in a way that made it as high as it could reasonably be. By making sure our background rate was on the high side, we could be confident that any modern rates we observed that were even higher were not just statistical artifacts but reliable indicators that extinction was proceeding faster than normal.

The next step was to look at the extinction rate for mammal species wiped out by people over the past few centuries to see if it was higher than the background rate. It was. In fact, over the past five hundred years, people have caused mammal species to go extinct sixteen times faster than normal, and over the past century, thirty-two times faster than normal. We also looked at other modern animal groups for which the IUCN had evaluated all of the species, namely amphibians and birds. Their extinction rates since people got into the act also were enormously high in comparison to the average mammal background rate: over the past century, the bird extinction rate is at least nineteen times the background rate, and the amphibian rate is as much as 97 times the background rate.

Those were sobering numbers. But they didn't answer the question of how the rate of extinctions caused by people in the past few centuries compared to extinction rates that prevailed during Earth's very rare Big Five Mass Extinctions. Besides the K-Pg extinctions, the Big Five included a mass extinction at the end of each of the following geological periods: the Ordovician (about 443 million years ago), the Devonian (359 million years ago), the Permian (252 million years ago), and the Triassic (200 million years ago). For three of these, species losses were even more dramatic than the 76 percent decimation seen at the K-Pg boundary: an estimated 86 percent of known species were lost in the Ordovician event, 96 percent in the Permian, and 80 percent in the Triassic (75 percent were lost in the Devonian). For our purposes, the interesting thing was the *rate* at which these species disappeared, in contrast to the final tally. Was it faster or slower than what is happening today?

That brought us up against another brick wall: how to resolve time to suitable levels of precision in the deep geological record. Because there are always error bars of hundreds of thousands of years for any dating method that can be used on ancient rocks, it is impossible to know exactly how many years elapsed from the beginning to the end of each of those extinction events. Nevertheless, mixing and matching various dating techniques makes it clear that each of the mass extinctions happened within a span of time that was between a few tens of thousands and two million years. Using those as bracketing numbers to plug into rate equations, in conjunction with the percentage of species estimated to die out at each mass extinction event, it quickly became evident that the rate of extinction that has prevailed over the past five hundred years is *faster* than one that would cause a mass extinction on a geological time scale. So much for my initial intuition that an apples-to-apples comparison would make today's extinction rates look less serious.

Our research group gained another perspective by assuming that each of the Big Five Mass Extinctions took only five hundred years to occur. We of course don't know how long those past extinctions took, but pretending they took only five hundred years was a way of understanding whether present rates could result in a geological-style mass extinction in a similar geological eye blink. The bottom line was that the rate of species loss needed to produce each of those past mass extinctions within only five hundred years was very close to the rates that would be evident if all of the IUCN threatened species actually went extinct. Put another way, continuing the species-loss trajectory we are now on will all but assure that we bring on the Sixth Mass Extinction in as little as three centuries. Lest you think that is a long time off, don't forget that this estimate assumes that human impacts will not increase. Those impacts certainly will increase markedly as we add a few more billion people to the planet, and the net effect could very well be to bring on mass extinction much sooner.

We published our results in the journal *Nature* in 2011.[16] Immediately the telephone began to ring and e-mails piled up as reporters looked for the right angle on the story. Some of them focused on the gloom and doom—everybody loves to read about a good cataclysm. But many news stories picked up on an even more important result of our study. We had also estimated how far we have actually come toward the Sixth Mass Extinction by tabulating the proportion of IUCN-evaluated species that people had killed over the past five centuries.

And there lay the glimmer of hope, the good news, if you will. For groups (like mammals and birds) for which the IUCN has evaluated 100 percent of known species, we have in fact lost less than 1 percent of species in historic times. If we add the species that have gone extinct due wholly or in part to human activities over the past 50,000 years, that percentage rises to about 7 percent for mammals and maybe 17 percent for birds.[17] If we also assume that the 122 amphibian species missing in action since 1980 are in fact extinct (the IUCN has not classified them as such yet), we are looking at a 3 percent loss for them.[18] Those kinds of numbers seem to be the case for most of the species we know much about. The losses of 1–17 percent of various species are unequivocally tragic, but compared to the 75 percent or greater loss that characterized the Big Five, we thankfully have a long way to go—in percentage-loss of species, anyway—before the Sixth Mass Extinction arrives.

Put bluntly, it's not too late. Those relatively low percentages of lost species mean we still have the chance to save the vast majority of life that has ridden the planet with us for the past 11,000 years, if we decide that we want to. But, as with so many things in life, timing is everything. Up to now we've been able to go about our business and leave the rest of the biosphere to take care of itself, but now we've reached a time where the already too-high extinction rates we've set into motion will very soon—within just a couple of human generations, if not before—accelerate out of control, if we don't figure out how to put on the brakes in very short order. Once we reach that point of no return, the outcome, just like the outcome of that asteroid hurtling toward Earth sixty-six million years ago, will be inevitable.

CHAPTER THREE

A Perfect Storm

This was no disaster movie. It was the real thing. In New York City, waterfalls rushed down the stairs into the subway stations, flooding the tracks for days afterward. The storm surge ripped up large sections of the Atlantic City boardwalk. Beachfront houses and shops, buffeted by the ocean waves and high winds, were turned into rubble in a matter of hours. A power station exploded, a flash of bright orange in the rain-soaked night sky, leaving millions of people without electricity for days, in some places weeks. In all, the damage in the United States spread through twenty-four states, covering the entire eastern seaboard and reaching inland as far as the Appalachian Mountains, Wisconsin, and Michigan. Losses were enormous—the weather catastrophe left more than 250 people dead and destroyed or damaged hundreds of thousands of homes and livelihoods. All said and done, dealing with the calamity was estimated to cost more than $65 billion. The disaster went down in history as Superstorm Sandy, though some called it a Frankenstorm because it brewed up around Halloween, hitting the northeastern coast of the United States on October 29, 2012.

Whatever the name, everybody agreed on one thing: it was a "perfect storm," so called because it took an unusual set of circumstances to make it happen. In Sandy's case, the unusual circumstances included a confluence of three, maybe four things. Sandy developed, like hurricanes typically do, in the tropics and began to move north. In the normal scheme of things, as hurricanes follow their northward trajectory they lose steam. Not so in Sandy's case. Arctic air coming down from the north actually intensified the storm as it skirted out of the tropics. At the same time, a high-pressure system was parked over the North Atlantic—that pushed Sandy right into the New York area. Most hurricanes never make landfall that far north. All of this happened during the full moon, which meant higher tides added more oomph to the already powerful storm surges. A fourth factor may have been the emerging effects of global warming: ocean temperatures have risen almost a degree Fahrenheit (0.5 degrees Celsius) over the past century, with about two-thirds of that rise happening in just the three decades from 1980 to 2012. That can increase storm intensity in two ways. First, warm water evaporates more quickly than cool water, putting more moisture into the air, which can cause storms like hurricanes to ramp up faster. Second, because warmer air over warmer oceans holds more moisture, rainfall can become considerably greater for any given storm.

"Perfect storms" come in many forms. Some are real weather events, like Sandy, or like the storm that twenty-one years earlier caught the *Andrea Gail* off the coast of Massachusetts, a tragedy that was later dramatized in a popular book and a movie.[1] Some are smaller in scale but no less important to the people they affect—like the unlucky coincidence of that driver who misses the stop sign at exactly the same time you enter the intersection, which in normal circumstances might work out ok, except the rainstorm that just started makes the road just a little slicker than normal ... And some perfect storms, and their drivers, are huge in scale—like the ways humans are changing the planet today.

If you are old enough to look back fifty years—or even thirty—those changes are pretty evident. Fifty years ago there were three billion people on the planet. Now there are more than seven billion. That's Change Number One. Each of those people has some basic needs—food, water, a place to live, and the energy required to produce those things. In order to serve those needs we have dramatically transformed nearly 40 percent of Earth's land surface into farms, ranches, cities, factories, and other human constructs. That's Change Number Two—a lot of suburbs, roads, and new agricultural fields have sprouted over the last five decades or so.

Change Number Three results from the way we have had to ramp up energy production to keep pace with the demands of an ever-growing human population—demands that include power generation, heat, transportation, and production of goods and services. We've achieved that by burning more and more fossil fuels—that is, oil and its derivatives (like gasoline, kerosene, and diesel), coal, and natural gas. Each barrel of oil or lump of coal burned releases a certain amount of greenhouse gases—mainly carbon dioxide, nitrous oxide, and methane—so called because when those gases accumulate in the atmosphere, they contribute to trapping some of the sun's heat that would otherwise radiate back into space. The net effect has been to raise the average temperature of the planet some 1.4 degrees Fahrenheit (0.8 degrees Celsius) in the past century, with most of the change occurring since 1950. That doesn't sound like much, but when you set it in the context of how much and how fast the climate normally changes on Earth, it turns out to be a lot. It's like the planet is beginning to run a fever relative to its "normal" operating temperature, which has been in place for all the time humans have been on Earth.[2]

The results are already evident most places—just talk to any old-timer (and by that, I mean somebody who can remember what was normal when they were a kid forty years ago or so), and you'll hear the local version of the story. The specifics of those stories vary from place to place because the climate system is complex, but the themes that keep cropping up are things like warmer winters (can't get out on the lake to ice fish, or the snow's lousy for skiing), spring arriving earlier (daffodils never used to bloom this early), different birds at the bird feeder (robins never used to be here in January), hotter summers (don't remember a June this hot and dry), and so on.

From a scientific perspective, we are seeing a world that on average not only is warmer than it was fifty years ago but also has more frequent extreme weather events—storms like Sandy that blow in from the ocean, or intense deluges inland that lead to major floods. We are also seeing many, many more record hot days versus record cold days per year (prior to the past couple of years, the number of record cold days for a year was about the same as the number of record hot days), longer than average droughts that are also more widespread, and longer than normal hot spells that are also hotter than normal—so much so that in 2013, Australia had to add two more colors to its weather maps to extend the upper end of the temperature scale. The oceans too are being affected: sea level has already risen over half a foot (roughly 0.17 meters) due to melting glaciers and the fact that warmer water expands, and ocean waters are becoming more acid as a result of the way seawater reacts with the higher concentrations of carbon dioxide in the atmosphere.

In short, as our numbers grow, we humans are spreading like ravenous army ants across Earth's surface, increasingly replacing the habitats other species need with our own specialized habitat, altering the chemistry of the atmosphere and oceans, and pushing the planet into a new climate state. Each of those changes has a huge influence on the survival of non-human species. Put them all together, and they amount to a perfect storm that is every bit as intense as—perhaps even more so than—the geological changes that caused each of the past five mass extinctions.

Recall that the dinosaur extinction I discussed in the preceding chapter was one of these, but famous as it is, it was by no means the worst. Wind the clock back 252 million years, and you are smack in the middle of something some paleontologists call the Great Dying. It's called that because a whopping 90 percent or more of Earth's species died out in a geological heartbeat, most likely within 200,000 years, maybe in as little as 20,000 years.[3] Even in that mother of all mass extinction events, the trigger involved only one piece of the perfect storm we now have forming.

The more staid name for the Great Dying is the Permo-Triassic, or P-T, mass extinction event, because it happened as the Permian period gave way to the Triassic. While there have been some scientific speculations that, like the dinosaur extinction event, the P-T Great Dying was at least in part precipitated by an asteroid impact, the evidence for that is at best very weak.[4] In contrast, an accumulating body of data indicates that the Great Dying was triggered by something that is all too familiar today: adding carbon dioxide—CO_2—to the atmosphere in unusually large quantities, which in turn changes the climate and oceans in ways that make it difficult for life on Earth to cope. At such times, the rules of the survival game change more rapidly than many species can deal with—the habitats that they have been evolving in for millions of years shrink dramatically or disappear altogether.

The end-Permian world, of course, was quite different from the world we live in now. For one thing, all of today's continents were more or less mashed together into a huge supercontinent called Pangaea. For another, the animals and plants of the time were strange by today's standards. There were no mammals or even dinosaurs yet, although the evolutionary branches that led to these more familiar animals had begun to sprout.

On the mammal branch were lots of odd-looking beasts called synapsids. Although at first glance you might, understandably, think they were some kind of mutant oversized lizard, if you took the time to examine their skulls (well, you'd have to get the meat off first), you'd see they had only a single opening for muscle attachments behind each eye, which aligns them with later mammals, rather than with reptiles, which have a double opening. That's where the name "synapsid," meaning "one-arch," comes from, in reference to the single arch of bone that forms the base of the lower part of the muscle-attachment hole in the head. Synapsids included animals like the gorgonopsids, whose name gives the right impression, coming as it does from the monstrous Gorgons of Greek mythology. Gorgonopsids were fierce carnivores of the end-Permian. They were about the size of large bears and sported canines that looked a bit like the slashing teeth of saber-toothed tigers, were backed by a row of big, sharp, piercing teeth, all the better to eat you with. Gorgonopsids were the ones that other synapsids, for instance *Dicynodon,* a chubby contemporary about four feet long and a couple of feet high, had to look out for. To see *Dicynodon*'s long tusks, you might think it, too, was a fierce predator, but in fact its flattened back teeth show it was, if not necessarily docile, a plant eater. Of course, there were many other kinds of four-legged critters as well, including the progenitors of the dinosaurs and of the animals we typically think of as today's reptiles, like crocodiles, lizards, and snakes.[5] Also around at the time were lots of insects, including a bunch of cockroach relatives, some big enough to give you nightmares if you are not an insect-loving person.

The plants might not strike you as too unfamiliar, if you happen to live in certain parts of the southern hemisphere. Forests were dominated by ferns, seed-ferns, and many kinds of archaic evergreen trees like ginkgos—probably giving very much the feeling one gets tromping through forests in New Zealand or on the west side of the southern Andes foothills. The species of plants that grew in those Permian forests, though, are long gone. We know them only by their fossilized leaves, fruits, and pollen, including, for example, the distinctive tongue-shaped leaves of *Glossopteris* trees. On the other hand, if you live in the northern hemisphere and think of oak, hickory, chestnut, maple, and so on when you think of forests, the Permian landscape would have been very strange indeed. Angiosperms, which include the vast majority of deciduous trees, bushes, shrubs, flowers, and grasses in most forests today, were totally absent and would not evolve for another 120 million years or so.

We know most about Permian life from seagoing animals, particularly those that lived along the shallow-water margins of the Pangaea supercontinent. A time-traveling snorkeler or diver would experience the disoriented feeling of being in a place that felt vaguely familiar but where nothing was quite what it seemed. To be sure, there were clams and snails not too different (in general form, anyway) from those we're familiar with today, but there were lots of brachiopods too. Brachiopods, which are rare now, are like clams in that there are two parts to their shell, but very unlike clams in that the two parts are grossly different from each other in size and shape. Instead of echinoderms, like starfish and sea urchins, there were crinoids, animals that looked like lilies attached to the sea floor, with stems and arms made out of stacked disks that look for all the world like strings of flat beads, each complete with a hole in the middle. Coral reefs abounded, but instead of being made of the colorful multi-shaped scleractinian corals that form the foundation for today's "rainforests of the seas" (so called because they support much of the ocean's species), the Permian reefs were built out of stout, compact, tabulate corals that looked a little like marine honeycombs, or rugose corals, which resembled clusters of horns-of-plenty jammed together with tentacles waving out of their openings. And stirring up the mud on the bottom would be a few scurrying trilobites, seagoing arthropods that looked a little like giant roly-poly bugs (or maybe you call them pill bugs).

All of these perished at the Great Dying. As Doug Erwin, a Smithsonian Institution paleontologist who has studied the P-T mass extinction, wrote, "Killing over 90 percent of the species in the oceans and about 70 percent of the vertebrate families on land is remarkably difficult."[6] In the nearly twenty years since Erwin wrote that, wonder at the carnage has not abated, but we've learned a little more about what brought it on.[7] It turns out it was not so difficult to kill so many species once CO_2 was pumped up to levels that were unusual with respect to what had been normal for the several million years prior to the extinction event.

That this happened is indicated by a variety of chemical signatures teased out of the geological record. Generally those chemical signals are obtained by looking for different isotopes of carbon and oxygen that were locked in sediments that later became rock or living things that later became fossils. Isotopes are different forms of the same atom—for instance, carbon can take the form of ^{12}C, ^{13}C, or ^{14}C, with the numbers referring to the number of protons plus neutrons that the element carbon (C) can contain. For oxygen (O), common isotopes are ^{16}O and ^{18}O.

The procedure for extracting isotopic signals that record the P-T extinction event involves collecting sequentially stacked samples of rocks or fossils as you work your way from the bottom to the top of a roadcut or cliff, then taking the samples back to the lab and putting a tiny bit of each into a mass spectrometer. The "mass" in the name comes from the fact that the spectrometer measures the masses and relative concentrations of atoms and molecules; thus it tells you the relative proportions of the different isotopes of carbon and oxygen (as well as of other elements, such as nitrogen, strontium, calcium, sulfur, etc.). With that information plugged into a few equations, it is possible to construct a profile of how carbon concentrations changed through time (from the carbon data) and to estimate the mean annual temperature of the oceans (from the oxygen data).

In the late 1980s, the German geologist Mordeckai Magaritz and his colleagues began to construct those profiles across the P-T boundary. Magaritz and his team traveled with their rock hammers and other assorted field gear to the southern Alps of Austria and Italy and hacked off samples of the Permian and Triassic rocks that are so nicely exposed there, carefully recording how far above or below the P-T boundary each of their samples was. Then they hauled their bags of rocks back to the lab and prepared them for the mass spec. When the data came out and the numbers were crunched, they saw that a major change in the ratio of $^{13}C/^{12}C$ in the sediments—expressed as $\delta^{13}C_{carb}$—coincided with the extinction event.[8] The change they observed is known in the isotope business as a "negative $\delta^{13}C_{carb}$ excursion," expressing that the amount of ^{13}C decreased relative to ^{12}C, and it indicated some complex environmental changes that Magaritz and crew thought spanned perhaps as much as a million years.[9] The Magaritz group suspected that the isotope excursion held clues to the extinction event, but the group's members were cautious in their interpretations, noting simply that their results might be a useful guide for future studies of the extinction-causing mechanism. A year later, a team led by William Holser, an Oregon geochemist who had coauthored the Magaritz study, began to refine the story further. Analyzing rocks from drill cores that spanned the P-T boundary in the Carnic Alps of Austria, the Holser team found that there was not just one isotope excursion, but multiple ones across the boundary. That convinced them that something big had happened at the P-T boundary, something that indicated global environmental changes.[10]

Following up on those intriguing clues, Jon Payne, now on the faculty of Stanford University, and his colleagues found another place where sediments had been laid down more or less continuously through the time the P-T extinction took place. This happened to be in ocean sediments that were subsequently thrust to the surface to form rock exposures in Guizhou Province in southern China. The precise sampling that Payne and his crew were able to accomplish not only confirmed the P-T carbon excursion in a different part of the world but also identified that the excursions repeated for millions of years into the Triassic, with some of the Triassic excursions being even more pronounced than the P-T boundary excursion.[11] Because $\delta^{13}C_{carb}$ values are the result of a series of chemical interactions that cycle carbon through the atmosphere, oceans, and sediments, the changes that Payne's team, and others before and after them, saw in the ocean carbonate meant that atmospheric CO_2 rose a lot just prior to and during the extinction event, and things did not settle down until well into the Triassic.

Just how much CO_2 rose is hard to pin down because the relationship between $\delta^{13}C_{carb}$ and atmospheric CO_2 is not a simple one, involving as it does the way carbon is transferred among its principal reservoirs—the air, living organisms, the oceans, soils, marine sediments, and rocks. Nevertheless, computer modeling of how those reservoirs trade carbon suggests that in order to produce the observed negative $\delta^{13}C_{carb}$ excursion, the late-Permian atmosphere must have gone from around 850 parts per million (ppm) of CO_2 just before the mass extinction to about 2,500 ppm by the time 90 percent or more of Earth's known species had disappeared forever.[12] In terms of gigatons of carbon (GtC) in the air, those numbers equate to a rise from about 1,810 GtC to 5,325 GtC, or a total rise of about 3,515 GtC. Other calculations suggest even more carbon would have to have been added to the atmosphere to cause the observed $\delta^{13}C_{carb}$ excursion—somewhere between 11,000 and 50,000 GtC.[13]

That's a tremendous amount of carbon. Where did it come from? Unlike today, there were no people around to pump CO_2 into the air. But there was something every bit as effective: volcanoes that erupted so much and so often that they make the volcanic eruptions humans have seen in historic times—like those of Mount Pinatubo in the Philippines, Mount St. Helens in the United States, and Chaitén in Chile—seem about as significant as popping a pimple. At the end of the Permian, vast volcanic fields in Siberia erupted. That's very clear from the remnants of huge lava fields, the Siberian Traps,[14] which now cover about 800,000 square miles (two million square kilometers), basically the middle third of Russia, a land area roughly equivalent to all of the countries in western Europe. Taking into account what has eroded over the past 252 million years, the volcanic fields probably originally covered an area three times that big.

When volcanoes erupt, they not only spew out lava and ash but also spit out many different gases, and one of those is CO_2. In the case of the Siberian Traps, calculations suggest that the magnitude of volcanism was enough to generate 11,000 to 30,000 GtC—well within the range of the estimates of the amount of CO_2 needed to generate the observed $\delta^{13}C_{carb}$ excursion.[15]

That rapid influx of CO_2 had the ultimate effect of warming the planet about 11 degrees Fahrenheit (6 degrees Celsius). In the context of today's world, that is an all-too-familiar number. It's very close to where we are headed by the year 2100 if we continue emitting greenhouse gases into the atmosphere at the pace we are now.[16] The Permian comparisons are not very encouraging when it comes to thinking about what increasing mean global temperature by that amount might mean in the way of extinction. The mean global temperature of the Permian world was probably not so different from that of today, despite the different continental configuration and the overall higher CO_2 concentrations in the atmosphere.[17] Mean global temperature for the end-Permian is estimated to be about 1.8 degrees Fahrenheit (1 degree Celsius) warmer than it has been for most of the twentieth century, and Permian tropical sea surface temperatures seem to have been similar to twentieth-century norms, between 75 and 86 degrees Fahrenheit (24–30 degrees Celsius).[18]

Estimating long-ago temperatures is of course not an entirely easy task, but again, information from isotopes trapped in the rock record makes it possible. In the case of Permian temperature, the relevant isotopes are those of oxygen, ^{16}O and ^{18}O, which are preserved in the microscopic tooth-like structures of long-extinct eel-like animals that were just a few inches long and were the progenitors of us and all other animals with backbones. Those tooth-like structures are called conodonts, and the animals in whose mouths they sat are called simply "conodont-bearing animals." That the toothy pieces have a name, while the animal that bore them does not (unless you want to get into scientific nomenclature), stems from the fact that for more than a hundred years after their initial discovery in 1856, what conodonts actually were was a total mystery, even though they were widely used by petroleum geologists as a way to figure out the relative age of rocks. The mystery was finally solved in 1969, when Montana geologist and paleontologist Bill Melton took then-student Jack Horner (the same guy who later became famous for his dinosaur discoveries) to dig fossil fish out of some rocks known as the Bear Gulch Limestone near the Little Snowy Mountains. As they were splitting rocks under the big Montana sky, Bill and Jack came across, in the words of Bill Melton, "a curious, carbonized impression of an animal that I could not identify in the field. Several others were found in the next five weeks."[19] Back at the lab, under the microscope, they saw embedded within the carbonized body impressions the tiny, jagged little fossils that for so long had been known as the mystery conodonts. Melton took the eight body impressions that held the conodonts to a meeting of paleontologists in Chicago, and word of what he had got around quickly. Again in Melton's words: "Dr. [Rainer] Zangerl looked at one of the conodont-bearing specimens and sent Dr. Ellis Yochelson down to find some of the conodont specialists in the audience. At no other time or place could so many conodont specialists have been summoned so quickly." The pronouncement was made by the leading conodont specialist of the time, Harold Scott: "I have just seen the conodont animal."

Conodonts are so common in certain rocks from the later Cambrian (about 505 million years ago) to the late Triassic (about 200 million years ago)—including rocks of Permian age—because they are made of the same mineral that builds your bones, apatite, which is mostly composed of decay-resistant calcium phosphate and thus easily fossilized. A key component of phosphate is oxygen. This is a lucky thing for scientists interested in learning what the temperature of the Permian was, because the oxygen that the conodont-bearing animals incorporated as they grew is essentially locked into the 252-million-year-old molecules in the apatite. Chinese paleontologist Yadong Sun and colleagues took advantage of that fact and used a mass spectrometer to determine the ratio of ^{16}O to ^{18}O in a series of conodont samples that spanned the P-T extinction event in the detailed rock record in Guangxi Province. It had long been known that the ratio of $^{16}O/^{18}O$ provided a reasonable paleothermometer for ocean temperatures, because the concentration of the two isotopes in seawater, and therefore in the body parts of organisms like conodonts that grow in seawater, is in large part a function of water temperature. Therefore, by determining the $^{16}O/^{18}O$ ratio in the conodonts, taking into account some other factors that control the ratio (such as amount of glacier ice), and plugging the results into equations that relate the ratio to ocean temperature, it is possible to track fairly closely the details of warming and cooling through long time periods.

What Sun's research group found—that just before the P-T extinction, tropical sea-surface temperatures were not too different from today's, and that they rose approximately 11 degrees Fahrenheit (6 degrees Celsius) from right before to right after the extinction—was consistent with model-based results and other evidence that had come earlier, but it provided a much more detailed picture.[20] Sun and his colleagues estimated that raising the temperature by approximately 11 degrees Fahrenheit made the tropics too hot for many organisms to live, and they found a distinctive signature of that in the post-extinction fossil record. For instance, fish fossils, marine reptiles, and calcareous algae, common as pre-extinction fossils, all but disappear in equatorial regions as the extinction event proceeds. They summed up their findings quite neatly in the title of their paper in the journal *Science:* "Lethally Hot Temperatures during the Early Triassic Greenhouse."[21]

Adding all that CO_2 to the atmosphere did more than just heat things up. The oceans absorb CO_2 from the air, and the net effect is to change the chemistry of seawater to make it more acidic. This was clearly a problem for the marine life of the late Permian, and it contributed in no small way to the extinction. It turns out that the ocean animals that were hit hardest were those whose physiology—that is, growth and development—would be most perturbed by ocean acidification. These included such animals as corals, brachiopods, and echinoderms.[22]

As with temperature rise, the ocean acidification comparisons are a bit disconcerting. Acidity is measured on the pH scale, with lower numbers indicating more acid waters. The pH of the oceans varies a little depending on where you are, but in general, surface waters had a pH value of about 8.13 or so in pre-industrial times, before we started to add substantial amounts of CO_2 and other greenhouse gases to the atmosphere. By 1989, in the subtropical North Pacific the pH was 8.12, and in 2009 it was 8.08. Equations predict that the pH of ocean surface water should fall about 0.15 pH units for every 150 ppm of CO_2 added to the atmosphere;[23] if CO_2 emissions stay at the level they're at now, we are headed for a pH of about 7.90 around the year 2100.[24] A change from a pH of 8.08 to 7.90 may not sound like a cause for concern, until you take into account that pH, much like earthquake intensity, is measured on a logarithmic scale, which means that each successive jump on the scale is a tenfold change from the previous number—and until you take into account that the pH of the oceans is estimated to have changed from around 8.10 to 7.90 at the same time as—you guessed it—the P-T extinction event.[25]

This is no small cause for concern with respect to extinction, as we are seeing already, both in nature and experimentally. Oysters in the Pacific Northwest of the United States began dying in 2009 because the waters had become too acid for the embryos to develop normally.[26] In aquariums, a variety of species placed in seawater that simulates the temperature and acidity of the near-future oceans do not fare well, if they survive at all. Among the problems are organ damage and death for inland silverside fish, a common species in estuaries of North America; severe damage to the liver, pancreas, kidney, eye, and gut within a month of hatching for Atlantic cod; and even learning disabilities in certain damselfish—they don't learn how to recognize the fish that commonly eat them.[27]

There is one more feature of the Permian oceans that bears a troubling resemblance to what's happening today—dead zones. These are large regions of ocean where the oxygen is literally sucked out of the water—such oxygen-depleted waters are called anoxic. Where oxygen depletion occurs, everything dies. Anoxic waters were widespread near the time of the P-T extinction, as is clear from various lines of evidence, including isotopic data from nitrogen, molecular fossils of a particular photosynthetic pigment made by sulfur bacteria (which can only live in oxygen-poor water), and biomarkers that tell a story of increased microbial biomass (again, characteristic of oxygen-poor waters) as the extinction unfolded.[28]

In today's world, similar anoxic zones have become increasingly widespread, most of them caused by pollutants flooding into the oceans. Major rivers carry elevated amounts of human-generated wastes, especially nitrogen that runs off from over-fertilized agricultural fields, and deposit it in the ocean, providing seasonal bursts of nutrients for ocean algae. With the nutrient influx, the algae rapidly multiply, then die, in such prodigious numbers that the chemical reactions that accompany their decomposition use up all of the available oxygen. Presto—dead zone. Dead zones have spread exponentially since the 1960s; by 2008 they were reported in more than 400 places, mostly along populated coastlines. The kicker: climate models suggest that warming temperatures will make existing dead zones worse and will add new ones.[29]

Earth has experienced relatively rapid changes in climate, ocean acidification, and dead zones at times other than the present and the P-T extinction. At least some combination of those three actors made an appearance in most of the Big Five Mass Extinctions.[30] The end-Ordovician and Late Devonian extinctions (which took place about 444 and 374 million years ago, respectively) coincided with anoxia and climate change. The end-Triassic (202 million years ago) extinction was contemporaneous with acidification and warming. Even the end-Cretaceous extinction (66 million years ago)—which, remember from the preceding chapter, featured an asteroid impact as its coup de grâce—was associated with both climate change and anoxia. It's also important to remember that the decimation of species that happened at each of those mass extinctions changed the world forever. Each time, it took hundreds of thousands, in some cases millions, of years for biodiversity to recover, and when it did, the kinds of animals and plants that dominated were very different from the ones that ruled Earth before the extinction event.

Now, you have to ask yourself: If the worst mass extinction of all resulted from the cascading effects of rising CO_2, climatic warming, ocean acidification, and dead zones; and all past mass extinctions seem to have involved at least two of those four factors; *and* we've got all four of those things not only going on today, but actually accelerating—what does that mean for the plausibility of another mass extinction? Furthermore, what does it mean when we throw into the mix the two other major ways that humans are affecting other species: taking over much of Earth's surface for our own needs, and relentlessly killing other species and destroying where they live for short-term economic gains?

What it means is that there is no doubt we are now brewing up the perfect storm to trigger the Sixth Mass Extinction. We're fueling three distinct and powerful drivers that are about to intersect as an extinction superstorm, the three I mentioned briefly in the preceding chapter—power, food, and money—each essential to human survival and comfort, but each taking an ever-growing toll on the survival of other species and on non-human-dominated ecosystems. In terms of power, the challenges we face are how to keep the lights on and cars on the road without disrupting our climate so drastically that the Permian-Triassic changes look mild by comparison, and how to exercise the power we have to choose the future. With regard to food, how do we feed ourselves and the two or three billion additional people guaranteed to be on the planet by the year 2050 without cutting so deeply into other species' habitats and numbers that most currently threatened species, and many more besides, disappear forever? As for money, how can we profit from natural capital—the many other species that ultimately form humanity's life-support systems—without spending down nature's principal so drastically that we go bankrupt? If we are to avoid the Sixth Mass Extinction, we have to change the business-as-usual model in all three of these arenas.

Making those changes will be no small or easy feat. But then, neither was abolishing slavery, or putting a man on the moon, or avoiding nuclear proliferation, or building a global communication system that connects billions of people with cell phones and internet access. One of many good things about the human species is that we are clever, resourceful, and pretty good at coming up with solutions when we recognize global threats. But of course, first we have to acknowledge the reality of the situation, and that's what the past three chapters have been about. The rest of the book is about how to put things on the right track—starting with power.

CHAPTER FOUR

Power

"Main thing to remember is you gotta keep your head down," Danny instructed. "If you don't, it's gonna get knocked off by the low rocks."

"Got it," I coolly said as I wedged my hard hat down a little tighter, made sure my miner's light was on, and shoved aside a few errant lumps of coal so I could scootch down a little lower. I was hoping my nonchalant response masked what I was really thinking, which was something along the lines of, "Great, I'm on the death train to the Cretaceous."

At the time I was working as a coal geologist for Atlantic Richfield Company, looking for new prospects in the hills of western Colorado, just below Grand Mesa. If all went as ARCO hoped, the coal we were finding would come on line in a couple of decades to help feed the world's ever-expanding need for energy. That morning two other geologists and I were sitting in the back cars of a coal train about to descend a couple of miles down into a mine near Somerset, nestled in a scenic valley just up the road from Paonia, where we lived in an idyllic farmhouse surrounded by peach, plum, cherry, and apple orchards. We wanted to get a three-dimensional view of the kinds of coal seams we had been discovering when drilling through the Cretaceous-age Mesa Verde Formation, stacks of sandstone, shale, and coal that formed the hills all around us. On that fall morning, the hills were covered with yellow and red scrub oak and, if you went far enough up Grand Mesa, with forests of aspen, their bright white trunks crowned by just-turning-gold leaves quivering in the sun.

Our ride was basically a big coal bucket on wheels, part of a scaled-down train that rolled in and out of the mine on narrow-gauge tracks. The coal cars carried miners in at the start of their shift; then the miners blasted, hammered, pried, and shoveled for eight hours or so, filling up the cars they rode in on and sending the train, now laden with coal, back to the surface. This went on over and over, for three shifts, twenty-four hours a day.

This, I pondered as we rumbled into the blackness (while remembering to keep my head down), was in no small way my history—coal mining was in my blood, as they say. My grandfather was a Colorado coal miner for forty years, just like those guys in front of me, going down at the start of the shift scrubbed clean, coming out black, and cheating death every day. He saw his share of men dying next to him, got carried out of at least one mine himself when it filled with smoke from a fire, and eventually died of what in those days was called "miner's asthma," more commonly known as black-lung disease or as the more neutral-sounding emphysema. It was that family lore that made me a little nervous about what I was doing that morning and that had also fixed in my mind that if I was going into the energy business, I was going to be the guy drilling down from up top, not the guy down in the mine.

All this is to say, by way of full disclosure, that I've got nothing against fossil fuels. I grew up in a town whose economic well-being depended on the coal coming out of the mines my grandfather worked in—that coal was essential to fueling the steel mills of the Colorado Fuel and Iron Corporation, where my other grandfather died on the job and where most of my uncles worked. The coal business landed me that ARCO job right out of college, giving me a good start on a career and the kind of adventures a guy in his twenties lives for—and believe me, it is an adventure to be pulling pipe on a drill rig, highest thing on the horizon, when a big lightning storm comes through. Even without the lightning, one mistake around that many tons of moving steel can cost a hand or a life, adventure enough in itself. The coal business also gave me my first-ever airplane ride—to an open-pit coal mine in Decker, Montana, as a field assistant in a U.S. Geological Survey engineering group tasked with figuring out exactly how steep you can make the wall of an open pit mine without it coming down on the miners below. And after that, the oil industry gave me a shot at the real money when I went to work on the Alaska pipeline, keeping one step ahead of the bulldozers as we surveyed the route for important archaeological sites. The pipeline snaked out behind us as fast as the welders could weld, a mammoth undertaking to get that newly discovered Prudhoe Bay crude flowing south where it could be of some use.

Fossil fuels have been good to me in other, more indirect ways too—the same ways they have been good to you and a few billion others. I like turning the lights on, and I like taking road trips. I like living in a house that is heated comfortably in winter and cooled sufficiently in summer. And I like my iPhone, my computer, all those big comforts and little gadgets that simply are not possible without a ready, cheap source of energy. And of course, I like food on my table and clean water coming out of my tap. And therein lies the problem, as far as the Sixth Mass Extinction goes. All those things take energy, and lots of it.

It's where we get that energy that is turning out to be a problem for other species. It comes from two sources: what is given to us, and what we produce. Up until about three hundred years ago, we, like all other species on Earth, pretty much had to make do with what was given to us. That gift comes from the sunshine that makes it to Earth's surface. Sunshine is, in fact, the ultimate energy, which you can measure in watts. Every square foot of Earth's surface in full sun receives a little over 100 watts of solar energy per hour. We think of watts in terms of light bulbs, so imagine screwing a 100-watt bulb into a patch of real estate about the size of a big flowerpot and watching it light up. Plants think in terms of photosynthesis, though, not watts. Plants are the primary producers in the food chain, along with algae and various bacteria. They receive energy in the form of light from the sun, and through photosynthetic reactions convert it into sugars and starches that the rest of us can use, and then promptly get eaten by somebody for their efforts. And then that somebody dies, decomposes, and returns some of that energy he or she stole right back to the plants as roots slurp nutrients from the soil. Of course, one of the ways the plant eaters may die is to get eaten by somebody else, which means, in essence, that before that energy gets back to the soil, it can be cycled through several species. Such is the energy flow through the global ecosystem, the global ecosystem being defined as all the world's species—including us—interacting with each other and with the nonliving parts of their environments—water, air, climate, and so on.

Because each living thing requires energy to stay alive, exactly how many organisms and species can exist on Earth is ultimately limited by how much energy there is and how evenly that energy gets distributed among individuals and species. Visualize the energy budget as a pie—the pie is only so big, but you can slice it in many different ways, from many equally sized pieces that feed lots of people (distributing energy among many species) to keeping a big slice for yourself and apportioning much smaller pieces to a few others (co-opting most of the energy for a single species).

The size of the solar energy pie is actually enormous—about 3,850,000 exajoules per year are absorbed by the atmosphere, oceans, and land.[1] To put that in context, in 2010 all of the energy the human race produced and used from all fossil fuels and renewable sources amounted to around 550 exajoules.[2] This means that every hour and fifteen minutes, enough sunshine energy hits Earth's surface to power all of humanity's activities for a year (at 2010 levels, anyway).

But in the absence of any intervening technology, only a tiny fraction of that incoming energy can be used to power living things, because of a conversion bottleneck that is very tight—energy in the form of light (sunshine) has to be converted into chemical energy by primary producers (plants, algae, and bacteria) and then passed up the food chain. That's where photosynthesis comes in: the primary producers use the sunlight to trigger a series of chemical reactions between carbon dioxide and water that eventually produce sugars and starches needed to power themselves and anything that eats them.

Ultimately, then, without help from technology the total amount of energy available to power the global ecosystem boils down not to how much sunshine strikes Earth, but to how much sunshine can be converted to chemical energy by primary producers. Primary production, which is typically expressed as net primary productivity, or NPP, is measured as the amount of carbon dioxide that vegetation takes in during photosynthesis minus the carbon dioxide plants release as they metabolize sugars and starches for energy. The unit of measurement is grams of carbon; estimates for the global amount of NPP on land range from about 46 petagrams (Pg) per year to 66.5 Pg, with the most recent comprehensive analysis settling on 53.1 Pg.[3] A petagram is 10^{15} grams; another way to look at it is one petagram equals one billion tons. This too can be converted to a measure of energy; using 53.1 Pg per year equates to using about 728 exajoules per year.[4] More energy is added from the primary production that occurs in the sea—a lot more—but it is less easily measured, so for now I'll just use what we know about land-based NPP to illustrate some points about how energy is distributed among terrestrial species.

The 728 exajoules per year (give or take) generated by photosynthesis seems to have been more than enough to power the global ecosystem on land, and all the species in it, before humans became so abundant on the planet. We know this because a big portion of all that NPP energy never got used—it ended up being buried and put into long-term storage, and by long-term I mean millions of years. The NPP energy that didn't get used on land accumulated as undecomposed plants in swamps, which eventually were deeply buried and metamorphosed into the vast coal seams that are found not only in those places where my grandfather and I worked, but in many other locations around the world.

In the oceans, the NPP that went to waste, as it were, included gazillions of marine plankton that didn't get scooped up by something bigger and fell into oxygen-poor waters (which prevented decomposition) at the bottom of the sea. The abundance of those one-celled plants and animals made the sea bottom mud rich in organics (essentially, carbon), which, when buried deep enough and under just the right conditions, were transformed into the underground oil reservoirs that we tap into today.

Things started to change after *Homo sapiens* made their first appearance around 160,000 years ago. At first, the balance of global energy, at least insofar as we can trace it on land, remained pretty much as it was before. We know that thanks to back-of-the-envelope calculations that estimate how many big-bodied terrestrial mammals, or megafauna, the NPP budget was able to support. Megafauna are mammals whose average body weight exceeds 100 pounds (44 kg); that is, animals that are sheep sized and bigger. Typically, megafauna aren't primary producers—they are instead consumers, meaning they eat plants to survive, or they are right at the top of the food chain, eating other animals.

"Food chain" is actually a bit of a misnomer. It's really more like a pyramid, with many species at the bottom and just a few at the top. At the bottom are many, many primary producers; one layer up, there are fewer first-level consumers (species that eat plants directly); at the next level, there are even fewer species of secondary consumers (species that eat the species that eat the plants); and at the top there are very few "top" consumers, predators like lions, tigers, bears, and us. Each of those layers—primary producer, first-order consumer, and so on—is called a trophic level. The reason you have fewer and fewer species in each trophic level as you progress to the top of the pyramid is that there is a net loss of the original energy that came from primary production as you transfer from one trophic level to the next. That makes sense when you think about it—part of the energy you get from something you eat is stored in your body as muscle, bone, fat, and so on, and that's the part of your energy budget that something that ate you would get. But most of the energy you get from food is used up by your everyday activities—that's the part of the original primary-production energy that disappears as you go from one trophic level to the next.

Megafauna, having big bodies they have to fuel, consume a lot; therefore it takes lots of primary producers to feed them. There is, in effect, a global carrying capacity for the number of megafauna bodies that can exist on Earth, which in prehistoric times was limited by the amount of energy that primary producers could produce, and by the way energy flowed through species in the various trophic levels. As it turns out, megafauna carrying capacity can be roughly tracked through time by using the fossil record to estimate how many big-bodied species there are on Earth.[5] There are equations that relate body weight to how many animals a square mile (or kilometer) can support—the bigger the animal, the more food it needs, and the wider it has to range to get that food. The same intuitive underpinning leads to equations that relate how large a geographic range you can expect an animal of a certain body weight to have. Finally, there are equations to relate the size of a certain fossilized part of a mammal—usually a molar tooth—to the average body weight of the animal. Put those all together, and it is possible to estimate the body size of an extinct species, how many animals there were on average in a square mile (or kilometer), and how many square miles (or kilometers) the entire geographic range of the species covered. From there it is easy to calculate a number (actually, a range of numbers) for the average global megafauna biomass—the total weight of all the individuals of all megafauna species.

I cranked out those numbers a few years ago, just for the fun of it. The idea was to figure out whether the growth of one very important megafauna species, *Homo sapiens,* was balanced by the loss of other megafauna species. After all, we're megafauna too, just like elephants, saber-toothed tigers, and elk—the average body weight of an adult person exceeds a hundred pounds. Estimating the weight of all the human bodies on Earth is actually a little easier than estimating the total weight of all the nonhuman bodies, because there are population-growth models that can be used that to some extent have been groundtruthed by the archaeological record. The final part of the megafauna calculation is to add in all the cows, pigs, sheep, horses, and other large animals that we grow to feed or otherwise serve us. (This was pointed out to me by my wife, Liz Hadly—who conveniently also happens to be a global-change scientist at Stanford University—as we were discussing the data over lunch at one of our favorite restaurants. I won't swear to it, but odds are good we were sharing a steak at the time.)

The results of my calculations surprised me and also frightened me a little bit. What emerged were two troublesome graphs.[6] One graph showed that the normal carrying capacity for megafauna on Earth was somewhere around 350 species. That number did not change for hundreds of thousands of years, until around 50,000 years ago, when it began to decline, at first ever so slowly. Then, around 13,000 to 10,000 years ago, megafauna species crashed to around 183—a number that has more or less persisted until the present day (although many of those big-bodied species have recently had their populations decimated and are classified as threatened with extinction). At the same time the megafauna crash occurred, the number of human bodies on Earth increased dramatically.

That probably wasn't a coincidence, from an energy perspective. Megafauna carrying capacity can be reached in two ways. One is the way that prevailed prior to that 50,000- to 10,000-year window—energy derived from NPP was shared more or less equally among many different megafauna species. The other way is what began to happen around 13,000 years ago: one species—us—co-opting an inordinately large part of the energy produced by NPP for ourselves as our numbers grew and we spread to all continents. Since the available NPP is a fixed sum, something's got to give. And what gave were other species—populations of many other megafauna collapsed, causing the species to which they belonged to quickly wither and die out. Once that collapse happened, the number of megafauna species stabilized, at much reduced numbers, for the 10,000 years leading up to recent times. But the global ecosystem had been fundamentally changed. Half the big-bodied species were gone forever, and the energy available to fuel megafauna had been rerouted so that more and more passed through a single big-bodied species—us.

The second graph was, if anything, even more troublesome. Instead of graphing numbers of species through time, it plotted total megafauna biomass through time—essentially, the total weight of all the big-bodied animals on Earth, humans and our livestock included. That graph showed megafauna biomass remained at pretty much the same level (around 200 million tons) through hundreds of thousands of years, then dipped precipitously (to less than 100 million tons) between 13,000 and 10,000 years ago, when so many megafauna species went extinct. The implication is that the nonhuman megafauna species went extinct more rapidly than humans (and their livestock) could multiply to fill the resulting biomass void. But fill the void we did, although it took 9,700 years. By three hundred years ago, megafauna biomass was finally back up to pre-crash levels, but with an important difference: instead of all that weight being distributed through many big-bodied species, an enormously large proportion of it was made up of human bodies and the livestock we breed.

Then something remarkable happened. Megafauna biomass suddenly began to skyrocket, in just three centuries rising to the level we see today—to nearly one-and-a-half *billion* tons—orders of magnitude higher that it was for all of the time humans have been on Earth. About one-fourth of the "extra" biomass is us; the other three-quarters are our livestock.

What that tells us is that somehow we broke through the primary-production energy barrier. No longer did the fixed sum of around 728 exajoules that could be produced from conversion of sunlight into chemical energy by photosynthesizers (on land) limit us. If you recall your history books, three hundred years ago corresponds pretty well with the onset of the Industrial Revolution, when we began to mine coal out of the ground in a big way. With that, humans became the first megafauna species to produce energy, rather than just waiting for it to filter up the trophic pyramid. The energy we began to produce, first from coal, later from oil and natural gas, came from fossil sunshine—fossil NPP, so to speak—that had lain in reserve for millions of years. Most of the fossil NPP we mine and pump out of the ground goes toward keeping people alive: for example, toward saving us from freezing to death in cold climates, growing and distributing our food (think of fueling tractors and combines, producing chemical fertilizers and pesticides, filling trucks with tomatoes), and pumping and purifying water for major cities. Not a small portion of it goes toward keeping us happy—driving to the soccer game, watching movies, playing Call of Duty.

Even though we add an enormous amount of energy that was formerly locked underground to the global ecosystem, what we add falls far short of fulfilling our total energy requirement, let alone helping other species much. We also have to dip deep into the primary energy that is generated the old-fashioned way, from photosynthesis—which, remember, is the only energy available to power other species' survival. In fact, we siphon off around a third (28.8 percent) of all of Earth's terrestrial NPP for ourselves; some estimates run to nearly 40 percent.[7]

The upshot is that in order to keep the human race running at the speed it did in 2010 takes about 211 exajoules from NPP and 550 exajoules mostly from fossil fuels, which adds up to 761 exajoules. Remember that the total amount of energy produced on land by NPP alone is on the order of 728 exajoules per year, so the human race currently needs more energy just for itself than has normally been available to power the entire terrestrial ecosystem.

That's one of the reasons the biomass graph scared me. It shouts out that the only reason we humans can exist in such high numbers—numbers that are far above Earth's normal carrying capacity for big land animals—is that we add a huge amount of energy to the global ecosystem, mostly through the extraction of fossil fuels. Without that extra energy, a lot of people would have to die, and the high quality of life that billions of people now enjoy and billions more aspire to would evaporate. It's as simple as that.

For other species, the picture isn't any brighter. After we take care of our energy needs, what's left for other species is only 516 exajoules per year—nearly one-third less than they have had to work with for the hundreds of thousands of years they have been on the planet with us (leaving aside the last three centuries). No wonder other species are starting to go extinct. The energy statistics I've been using are only for the land, remember, but that doesn't mean the same general principles aren't in play in the sea. We're also starting to see extinctions in marine settings as we take more and more of the oceans' net primary productivity for ourselves.

It's not just that we're commandeering a big chunk of other species' NPP *and* have to produce more than twice that amount of energy in order to keep society ticking that made me sit up and take notice when I was thinking about those biomass graphs, though those issues were worrisome enough. Where the Sixth Mass Extinction is concerned, the most troubling aspect of the energy equation is that in order to provide the additional power the global ecosystem now needs, we rely primarily on fossil fuels. The problem, of course, is that burning coal, oil, and natural gas at the pace we now do puts tremendous amounts of CO_2 and other greenhouse gases into the atmosphere. As we saw in the preceding chapter, the resulting climate disruption alone could plausibly be capable of triggering a mass extinction on a par with the Permian Great Dying. And that doesn't even take into account that the interactions of climate change with other human-caused extinction pressures could make things even worse (which I'll talk about in later chapters).

So therein lies the energy conundrum. How do we keep producing enough energy to avoid a severe human population crash and keep quality of life at least as good as it is now, without changing our climate faster than other species can cope with? The answer is crystal clear: we have to replace fossil fuels with something else. And fast.

Just how fast becomes apparent with a little simple math, using a measure that is relatively straightforward to conceptualize—the number of barrels of oil you burn through in a year.[8] On average, as of 2011 each person in the world used 4.6 barrels of oil each year. Keep in mind, of course, that usage is much higher in some countries, much lower in others—for instance, per capita usage is 81 barrels per year in Singapore, 39 barrels in Saudi Arabia, 22 barrels in the United States, 10 barrels in France, 5 barrels in Brazil, 3 barrels in China, 1 barrel in India, and one-fifth of a barrel in Bangladesh and Nepal. Each barrel of oil that is burned releases about 0.43 tons of CO_2 (equivalent to 0.117 tons of carbon),[9] and each additional 500 to 600 billion tons (approximately) of carbon in the atmosphere warms mean global temperature about 1.8 degrees Fahrenheit (1 degree Celsius).[10] Multiply seven billion people by 4.6 barrels of oil by 0.43 tons of CO_2 per barrel, and you get the tons of CO_2 emitted in one year. Then grow human population for a year by the amount demographic models indicate, multiply it out again, add it to the last year, and repeat.[11] The numbers get very big very fast.

Let's say we get a little more efficient in our use of fossils fuels, and we burn on average only four barrels per person per year for the next century. By the year 2100—a scant human lifetime away—the resulting emissions would raise average global temperature about 1.8 degrees Fahrenheit (1 degree Celsius—and that's just from burning oil.[12] Add the emissions from burning the coal and natural gas[13] we also rely on and you raise the temperature in the neighborhood of 7.2 degrees Fahrenheit (4 degrees Celsius) within the lifetime of younger people alive today. And remember, that's assuming the world actually *reduces* oil consumption. That would require that those of us who use twenty-two barrels of oil per year (or any amount more than four) really cut back, to balance things out as those people who use less than four barrels (by far most of the people in the world, including those in China and India) begin to increase their usage. Otherwise, temperature rise goes even higher. The models that assume we'll continue to follow the approximate trajectory of fossil fuel use we've been on for the past decade indicate that by 2100, there is a 66 percent chance that Earth will be from 8.6 degrees Fahrenheit (4.8 degrees Celsius) to as much as

11.5 degrees Fahrenheit (6.4 degrees Celsius) warmer.[14] In other words, a global temperature increase that approaches or even exceeds that of the Permian Great Dying is very plausible—and remember that at that mass extinction, the heating took place over many thousands of years, not just a few decades. Carry the calculations out past the year 2100, and things get really grim.

Recognizing that business-as-usual energy production and consumption will almost certainly trigger the Sixth Mass Extinction is one thing. Doing something about it is something else entirely. In order to avoid the worst impacts of climate disruption, we need to reduce carbon emissions to near zero by mid-century, around 2050.[15] I'll say right up front that there is no silver bullet for hitting that target. Luckily, though, there is a shotgun. If we mix and match a variety of new technologies and tried-and-true approaches, feasible solutions emerge—if the incentives to implement them are strong enough.

Hitting the zero-emissions target by 2050 has three phases: (1) slowing growth of emissions, (2) stopping growth of emissions, and (3) reducing emissions.[16] The trick is to get to that third phase—actually reducing emissions—as fast as possible. The shotgun pellets (to continue that metaphor) come in several different guises, which ultimately take the form of using fossil fuels more efficiently in the wind-down to zero emissions while simultaneously rapidly ramping up and phasing in carbon-neutral energy-production technologies. Smart people—many of them entrepreneurs who stand to make billions of dollars off of the energy revolution—have been thinking hard about how to do this in recent years. One of the big breakthroughs in highlighting the feasibility of slowing and then stopping growth of emissions (phases 1 and 2) came from two Princeton University scientists, ecologist Stephen Pacala and physicist Robert Socolow. In an important paper published in the journal *Science* in 2004, they pointed out at least fifteen ways we can reduce emissions, each of which could keep 25 billion tons of carbon (25 GtC) out of the atmosphere over the ensuing fifty years.[17] Most of these are things we already know how to do. By picking just seven of those multiple potential solutions and working hard to implement them, they contended, it would be possible to keep 175 billion tons of carbon out of the atmosphere over fifty years. This, they calculated, would on average hold yearly carbon emissions to the levels characteristic of the early 2000s while accommodating the needs of more and more people—an outcome far better than having emissions rise year by year in lockstep with population growth, as has been the case for the past sixty years. Fifty years into the future, carbon-neutral energy technologies should have progressed far enough to begin reducing carbon emissions by 2 percent annually (phase 3), until we hit zero emissions. That strategy, Pacala and Socolow contended, should stabilize climatic warming to about 3.6 degrees Fahrenheit (2 degrees Celsius), which is probably about the best we can hope for given how much we've already heated up the planet and the lag time it takes for greenhouse gases to leave the atmosphere. From the extinctions perspective, a 3.6-degree Fahrenheit warming is not ideal, but it probably would be livable for many threatened species, particularly if we help them out a little (as I'll discuss in a later chapter). After all, we've already warmed the planet almost half that much, mostly since 1950, and although we are certainly seeing lots of dangerous trends in species reactions, we have not witnessed actual extinctions from climate change alone yet (although we've seen plenty from other human impacts).[18]

Pacala and Socolow called each of their solutions a "wedge," in reference to the wedge shape of the graph that results from plotting the accumulation of carbon emissions through time (or conversely, by plotting the amount of carbon emissions that can be avoided through time by implementing carbon-cutting solutions). They specified fifteen solution wedges that fall into five general categories: using fossil fuels more efficiently; making power-generating facilities that run on fossil fuel more efficient; using fossil fuels but capturing the carbon emissions so they don't get into the atmosphere (a technique called carbon capture and storage); replacing energy generated by fossil fuels with carbon-neutral energy production; and practicing more prudent agriculture and forestry. They suggested that by implementing any combination of seven of the fifteen solutions, we could keep enough carbon out of the atmosphere to prevent the worst climate-change impacts. Here is a summary of their list, which they were quick to point out is not exhaustive.

Use fossil fuels more efficiently. (1) Double the fuel efficiency of cars. (2) Halve the average distance each car owner drives per year, from the current 10,000 miles to 5,000. (3) Cut electricity use in homes, offices, and stores by 25 percent.

Increase efficiency in power-generating facilities that run on fossil fuels. (4) Raise operating efficiency at 1,600 coal-fired power plants from 40 to 60 percent. (5) Replace 1,400 coal-fired power plants with natural gas–fired plants.

Implement carbon capture and storage (CCS). (6) Install CCS at 800 large coal-fired plants. (7) Install CCS at coal plants that would produce hydrogen for 1.5 billion hydrogen fuel cell–powered vehicles. (8) Install CCS at coal-to-syngas plants (these convert coal to synthetic "natural" gas, then use the more efficient syngas to generate power).

Replace energy produced by fossil fuels with energy generated by carbon-neutral technologies.
(9) Add twice the nuclear-power generating capability that was available in 2004. (10) Increase wind power 40-fold to displace coal. (11) Increase solar power 700-fold to displace coal.
(12) Increase wind power 80-fold to produce hydrogen for cars.
(13) Drive two billion cars on biofuels.

Practice more prudent agriculture and forestry. (14) Stop all deforestation. (15) Expand conservation tillage to 100 percent of the world's croplands.

Clearly, some of these solutions would be easier and some more desirable than others in the long run. The ones that would invest a lot into jury-rigging temporary fixes for fossil fuel technologies could be money down the drain, as carbon emissions ultimately have to disappear to solve the problem. Other approaches, like innovations that are leading to better electric cars and energy-storage batteries, are not mentioned (I'll discuss some of those later). Nevertheless, those fifteen solution wedges that Pacala and Socolow highlighted nicely illustrate that we're not helpless; if we want to, we can start fixing things today. Of course, if more than seven solution wedges were implemented, the carbon savings would be correspondingly greater.

Which is lucky, because now it looks like we will need more than seven wedges to pry us out of the trouble we are in. A recent study by Steven J. Davis and colleagues, published in 2013, revisited the carbon-wedge concept and revealed that fixing the carbon problem will now require even more aggressive and rapid action than Pacala and Socolow thought.[19] Nearly ten years after the original wedge study, we know that at the global level we didn't start cutting emissions nearly enough in 2004. Numbers compiled and published in 2012 in the PricewaterhouseCoopers Low Carbon Economy Index show that emissions reduction from the year 2000 to 2010 averaged 0.8 percent globally.[20] In order to not exceed that "safe" 3.6-degree Fahrenheit temperature rise that Pacala and Socolow (and the rest of the climate-science community) were shooting for, we now have to reduce yearly emissions an average of 5.1 percent per year each year until at least 2050. That is going to be tough—the world has not seen that rate of reduction in any of the past fifty years. But we haven't really been trying too hard. Actually, we've hardly been trying at all.

What if we do try hard? What are the prospects for success? Reasonable, it seems, at least from a technological perspective, based on the recent work that has been going on in the energy sector. In the words of energy czars Steven Chu and Arun Majumdar, there are in fact "several research and development opportunities and pathways that could lead to a prosperous, sustainable and secure energy future for the world."[21] The quote is from their 2012 article on emerging energy technologies. At the time it was published, Chu was the United States Secretary of Energy and Majumdar was the Director of the United States Research

Projects Agency–Energy. The opportunities and pathways they laid out all involve doing what humans do best—inventing new things and then scaling up production to make the new inventions profitable. Chu and Majumdar were not talking pie in the sky; they were talking about breakthroughs that are already in the works, some on the cusp of scaling up, that will make it feasible to replace carbon-emitting power generation and transportation with carbon-neutral power.

Which takes us back to some of those fifteen solutions that Stephen Pacala and Robert Socolow suggested a decade ago. Do they really stand up from the technological perspective? Ten years' worth of innovation, engineering, and actual use of new technologies are providing some answers. First, those solution wedges that concentrate on increasing efficiency in the production and use of energy from fossil fuels are going to be important. It's become very clear that efficiency gains—in transportation, buildings, and fossil-fuel power plants—will provide the quickest and most easily achievable carbon cutbacks in the short term, that is, the next couple of decades.[22] This is because it will take at least one generation of "sustained efforts by both the public and the private sectors" to transform the existing energy system into one where carbon-neutral technologies dominate—you can't replace the whole energy system overnight.[23] But that is all the more reason to start changing it sooner rather than later. Every year of delay makes the problem worse and ends up costing governments, taxpayers, and the private sector more money in the long run.[24]

So what to do? The energy system divides into two fairly separate parts, the transportation system (powering vehicles that move us and goods around) and the stationary system (power plants that deliver electricity and natural gas to your house). If you thumb back a couple of pages, you'll see that three of the original carbon-saving wedges dealt with technological breakthroughs in the transportation system—better gas mileage for cars, more use of biofuels, and entirely new fuel systems, like hydrogen. All of these are beginning to come online. Gas mileage is almost certain to increase rapidly thanks to technological advances that make cars lighter, reduce friction in moving parts, convert waste heat into energy, and improve aerodynamics. Just making cars lighter by using advanced steels and carbon-fiber-reinforced composite materials results in huge energy savings: every 10 percent loss of weight improves fuel consumption by 6 to 8 percent, and in the next ten to twenty years, it seems likely that we will be able to reduce vehicle weights by up to 40 percent while still making cars that are safe to drive.[25] Given that new cars sold in the United States in early 2013 averaged 23.5 miles per gallon, weight reduction alone could mean that the average new car in 2033 would easily get over 30 miles per gallon.[26] Another 15 percent improvement can be expected from innovations under way to increase the efficiency of internal combustion engines; that kicks average miles per gallon up to about 35. Even further fuel savings accrue with gas-electric hybrids that are already routinely going 40 to 50 miles on a gallon of gas. Indeed, miles-per-gallon standards in many countries are already significantly higher than the average for the United States; for instance, in Japan and Europe the requirement is 44 miles per gallon.[27]

The bottom line is that there are no technological impediments to doubling the average car's gasoline mileage within a few decades, a reality that is already being acted upon. With the support of both industry and environmental groups, new fuel efficiency standards were emplaced in the United States in August 2012. Those standards require an average of 34.5 miles per gallon by the year 2016 and 54.5 miles per gallon by the year 2025 (this latter phase of the standards applies to model years from 2017 to 2025). Enacting those standards will essentially double gas efficiency compared with that of vehicles manufactured in 2008. This reflects a significant change in people's thinking about what is possible and desirable: the Energy Independence and Security Act of 2007 viewed increasing then-existing standards to just 35 miles per gallon by 2020 as a major achievement.

Applied worldwide, doubling fuel efficiency could halve a major part of the transportation system's carbon footprint in about a decade.[28] Is that all we have to do to get out of the woods? A few quick calculations give you some idea. In 2010, cars were responsible for emitting about 4.4 gigatons of CO_2 per year.[29] Holding the number of cars steady and not changing current average gas mileage for the next fifty years would mean 296 gigatons of CO_2 would be emitted.[30] That's way too much. No more than 71 gigatons can spew out of vehicle exhaust pipes from 2015 to 2065 if the goal is to reduce vehicle emissions on average by 5.1 percent per year for the next fifty years (remember, a 5.1 percent per year reduction is what we need to hold to a workable temperature rise of 3.6 degrees Fahrenheit or 2 degrees Celsius). What if we double car mileage worldwide by 2025? We still end up adding 186 gigatons from vehicle exhausts by 2065, roughly 115 gigatons too much.[31]

Even though these estimates don't come near to hitting the emissions-reductions targets, they are probably overly optimistic because they hold the number of cars steady. In reality, there will likely be many more cars in the world by 2065. Not only will there be more people (therefore the number of cars will grow if we hold the ratio of cars per person constant), but it is also likely that rising affluence will lead to more people buying cars, so the average number of cars per person would rise too. What happens if we take that growth into account? We can start with the year 2009, when there were about 134 cars per 1,000 people worldwide. As we saw with barrels of oil used per person, the global average by no means implies that all countries have 134 cars per 1,000 people. For example, in the United States there are 627 cars for every 1,000 people, versus 2 cars per 1,000 people in Bangladesh and Rwanda.[32]

Let's assume that world population grows to ten billion before it stabilizes in 2050, and that more and more people can afford to own cars.[33] A middle-of-the-road (so to speak) scenario is to grow the world average to about 214 cars per 1,000 people over the next 45 years—about what it was in 2010 in Serbia—and then stabilize at that number.[34] In that case, there would be about two billion cars on the streets by 2055; presently, the world has about one billion. Assuming that by 2025 gas mileage will be twice as good as it is now, by 2065 the cumulative exhausts would have emitted 251 gigatons of CO_2. That's a whopping 180 gigatons above the maximum of 71 we can allow in the road transportation sector if we are to meet emissions reductions of 5.1 percent per year over the next five decades.

You are probably beginning to get the message: simply doing what we've always done with fossil fuels, but doing less of it, would certainly help, but it is just not going to accomplish the job. We've already missed that boat. The only thing that *will* do the job, if we want to stick with some version of the internal combustion engine, is to use a fuel other than a fossil fuel.

Which takes us to biofuels. You very likely already run your car on some biofuel—ethanol additives, which on average make up a little more than 5 percent of the gasoline that goes into many people's tanks throughout the world, is produced from crops such as corn, sugarcane, soy, and palm oil plants. While biofuels like ethanol are a little less efficient than gasoline in terms of power produced, they have the advantage of leaving a much lower carbon footprint if they are produced from the right crops. This is because, unlike traditional gasoline, biofuels do not release *fossil* carbon when they burn. Instead, they emit only the carbon that was gulped out of the atmosphere by the crop that was grown to produce the biofuel.[35] Ideally, biofuel production and use can approach being a carbon-neutral cycle—plants gulp in carbon while growing and release carbon when burned, then the next generation of plants gulps it down again. Biofuels also have the advantage of letting us keep doing things that would become very, very difficult if gasoline simply weren't an option—like flying airplanes. That fact has brought the armed forces into the act. The U.S. Navy and Air Force are aggressively working toward bringing biofuels online as part of overall U.S. national security strategy. As an example, in 2011 the departments of the Navy, Energy, and Agriculture teamed up on a $510 million, three-year joint venture with the private sector "to develop advanced biofuels compatible with existing military infrastructure."[36]

Say what you will, once the military sets its mind on technological breakthroughs, they tend to happen. In fact, some already have. The Air Force has flown F-15 and F-16 fighter planes and C-17 transport planes on 50 percent biofuel, and the Navy successfully powered a carrier strike group on biofuels for a day to demonstrate that it would work.[37]

Two things stand in the way of replacing traditional gasoline and jet fuel with biofuels: cost, and the conflict between using agricultural land to feed people versus feeding cars and planes (I'll discuss this latter point a bit more in the next chapter). In the past few years, however, a field of research called synthetic biology has been ramping up and providing discoveries that are capable of breaking down both of those barriers.

Basically, synthetic biology is using genetic engineering to build life forms that will help us out. It's a bit scary and a bit weird to think about, but it's happening already—a classic example being "spider-goats." These are goats into which geneticists have inserted a gene from spiders, the very gene that gives spiders the ability to spin the silk for their webs. The genetically altered goats secrete spider silk in their milk; the silk can be strained out and used to make substances that require a tensile strength greater than steel's.[38]

That sort of biotechnology is now being applied in the biofuels industry, where it gets around the food-versus-fuel problem by using algae or microbes for the biofuel source. For instance, whereas replacing all the petroleum used in the United States by biofuels produced from soy or palm oil plants would require converting 330 percent or 23 percent, respectively, of the land area of the United States to production of those crops, algae could do the job using only 4 percent of the U.S. land area.[39]

So far, the problem with algae-based biofuels has been one of production costs, which remain prohibitively high, and that's where genetic engineering is coming into the picture. A key part of the cost is microbial productivity—how many algae or other microorganisms can be harvested in each growth cycle. Helpful genetic tweaks that have already been discovered to enhance productivity include those that increase triglyceride concentrations (triglycerides are a key compound for producing oil from the algae), growth rates, and cell density. Metabolic engineering and synthetic biology also are being used with promising results to explore biofuel production from microbes other than algae, including yeast, bacteria, and microbes that convert sunlight into hydrocarbons without using photosynthesis.[40]

Even algae and microbes have to eat, so also important to the ultimate cost of biofuels from such microorganisms is the source of their feedstock. Feeding them simple sugars and starches is efficient but expensive, and it also has an upstream carbon footprint, the one generated by producing the sugars and starches. The workaround, where much attention is currently being focused, is using waste from agriculture and the wood industry as feedstock. This requires converting lignocelluloses into something more readily digested by the biofuel-producing microorganisms. Other innovative and imminently feasible solutions include farming algae near nutrient-rich wastewater streams or sewage plants and diverting that nutrient-rich wastewater to feed the algae ponds.[41] Recall from the preceding chapter that the excess nutrients being dumped into the oceans from such sources are already stimulating algal blooms that are causing dead zones. It seems a no-brainer to harness those nutrients for something useful and keep the oceans alive.

Algae-based biofuels are more than a potential for the future. Factories are already producing them, and venture capitalists and energy companies have started funneling hundreds of millions of dollars toward research and development of them. ExxonMobil recently teamed up with Synthetic Genomics to the tune of $600 million to improve algae fuel extraction and growth techniques. Cascade Investment (of Bill Gates fame) and Venrock (of Rockefeller fame) invested tens of millions of dollars in Sapphire Energy, one of the leading algae research companies, which also happened to be listed by Forbes as a company to watch (as of 2014).[42] No wonder—Sapphire has already made algal jet fuel, which worked just fine in the planes that Continental Airlines used to test it.

Scaling up suitable biofuels to make them economically viable is not the only way to significantly decrease the carbon footprint of the transportation sector. The other way is to get rid of the internal combustion engine altogether. It is, after all, a pretty antiquated technology—how many other things that were invented over 150 years ago are still considered state of the art? Alternatives to the internal combustion engine are already off the drawing boards in the form of fully electric vehicles, which use no liquid fuels and come in two varieties. One variety uses hydrogen-fuel cells: hydrogen gas is used to generate electricity, which in turn powers the vehicle. Experimental models have proven the concept and work is now progressing to remove a key cost impediment, namely that these vehicles need an on-board tank that will hold the pressurized hydrogen gas safely even in the event of a collision. Currently those tanks cost about $3,000 a pop, but advanced materials and production techniques will likely bring costs down, especially if production is increased.[43] Another stumbling block for the short term is how to deliver the hydrogen to refueling stations—that is solvable, of course. We have had no problem building similarly large and complex delivery systems for oil, gasoline, and natural gas, as I saw firsthand with construction of the Alaska pipeline, and as you've probably witnessed when a road crew tears up the street to replace a water main or gas line. Another hurdle is finding a way to produce the amount of hydrogen needed without adding to the carbon footprint—which is the reason for having electric vehicles in the first place. This, however, can potentially be solved by such techniques as using solar or wind power to produce hydrogen from water, essentially a carbon-free process. Similarly, generation of hydrogen from water can be achieved by farming cyanobacteria, which produce hydrogen through their normal metabolic functions.[44]

The second kind of electric vehicles are plug-in models, so named because they are plugged into an electrical socket to charge their batteries, with the socket of course ultimately getting electricity from a power-generating station. A variety of plug-in vehicles—from the fairly affordable Nissan Leaf to the high-end sports car made by Tesla—already are on the road, and they will likely get cheaper and become more common as battery technology advances. The amount that plug-in electric cars reduce emissions relative to gasoline-powered cars depends on the source of the electricity used to charge them. If cars are charged with electricity that comes from a coal-fired plant, they may have a carbon footprint that is equivalent to 33 miles per gallon of gas, but if the electricity comes from cleaner plants—gas fired, nuclear, hydroelectric, solar, or wind—the carbon footprint can equate to at least 79 miles per gallon.[45]

This brings up an interesting point about fully electric cars. Unlike with gasoline-powered cars, the fuel-efficiency gains of fully electric vehicles become tightly coupled to the stationary part of the energy system. The stationary energy system includes the electricity-generating plants (and natural gas plants) that today are powered by coal, natural gas, water rushing through a dam turbine, nuclear reactions, windmills, or solar power captured by photovoltaic or concentrating cells.

Which brings us back to those coal mines that I and my grandfather got our start with. Coal is really the path of least resistance—the old ways of digging it out of the ground still work, it's abundant and relatively cheap, and it has high bang for the buck in terms of energy production. Unfortunately, it is also just about the worst thing you can do in terms of putting excess greenhouse gases into the atmosphere. Coal emits about 205 to 227 pounds of CO_2 for every million Btus (British thermal units) of energy it produces, compared to 161 pounds for heating oil or diesel and 117 for natural gas.[46] If we continue using coal at the per capita rate we did in 2010, by 2065 we will have released about 996 gigatons of CO_2 (which equals 271 GtC).[47] That's enough, all by itself, to raise global temperature another degree Fahrenheit or so (0.5 degrees Celsius) in just fifty years. A reduction in coal emissions by 5.1 percent per year for the next fifty years—remember, that is the across-the-board reduction we need to stabilize the temperature rise from all fossil fuel emissions below 3.6 degrees Fahrenheit (2 degrees Celsius) by the year 2100—would mean we have to get coal emissions down to 279 gigatons of CO_2 (76 GtC) by 2065.

How do we avoid burning coal in order to accomplish that much reduction in five decades? There is a lot of talk about substituting natural gas for coal, since the emissions from burning natural gas are about half of those produced by coal.[48] That prospect, in fact, has stimulated a gas rush, especially now that the relatively new technology of fracking makes it economically tractable to retrieve gas and oil from otherwise "tight" rock, like shale.[49] The truth is, substituting natural gas for coal is a Band-Aid at best, and not a very effective one. Even if we were to replace every coal-burning facility in the world with a natural-gas facility, phased in gradually from now until 2030, at current per capita energy consumption the result would be about 580 gigatons of CO_2 (158 GtC) emitted by combined gas and coal by 2065. That's more than twice as much (301 gigatons of CO_2 too much) as our reductions target allows.

Carbon capture and storage, or CCS, is also often touted as a viable solution, but so far it faces some pretty severe obstacles to deployment on large scales. CCS is a technological fix whereby the CO_2 emitted from burning fossil fuels is captured and diverted somewhere it will do no harm, usually far underground. The problem is that retrofitting existing coal plants with carbon capture systems presently costs as much as building a new power plant, and then 20 percent to 40 percent of the plant's energy-producing capacity has to be diverted toward separation, compression, and transmission of the CO_2.[50] This means you have to find another way to replace that lost power (hopefully not by building another coal plant). Moreover, the technology and feasibility are far from proven on an industrial scale. Methods would have to be "developed, piloted and demonstrated at near-commercial scale before wide-scale deployment can occur."[51] Although recent studies suggest that carbon capture and storage, if combined with carbon-neutral biomass-powered facilities to provide the extra energy needed to fuel the process, could be helpful in reducing carbon emissions by 2150 if deployed widely by 2050, in the short run, CCS alone will not free us from our carbon emissions problems.[52]

For the stationary energy system, then, the story is very much the same as in the transportation system—the only way out is to replace fossil fuels with something else. For electricity generation, we have that "something else" in spades. I've already pointed out that one hour of incoming solar energy is enough to power the world for a year. Likewise, recent work has demonstrated that wind can supply as much as several times the world's all-purpose power needs.[53] The technology is already in place for both, but wind and solar combined, as of the time I write this, account for under 3 percent of the world's power generation.

Other sources of carbon-free energy include hydroelectric, geothermal, and nuclear power. These are presently critical components in the stationary energy system, although they remain minor pieces in comparison to fossil fuels. Hydroelectric power is what we get from water driving turbines, and geothermal energy comes from tapping into the hot water that exists at drillable depths in many parts of Earth's crust. Nuclear power is extremely efficient and clean but has obvious downsides in the form of potentially devastating accidents and problems with waste disposal. Nevertheless, proven nuclear technology, especially safer next-generation plants, can provide an effective building block in the transition to a low-carbon energy system.[54] Another potentially huge source is tidal and wave energy—using the ocean's tides and waves to generate power.[55] Such systems are just starting to be employed in a few places.

In principle, there is no reason that carbon-neutral energy production cannot be scaled up dramatically and quickly—and it's beginning to happen. In 2009, solar installations were growing at 20 percent per year, and for wind, emerging markets are expected to drive growth of at least 8 percent per year until 2016.[56] In fact, we have the capability to scale up existing technology much faster, as recent work by climate and energy researchers Mark Jacobson, at Stanford University, and Mark

Delucchi, at the University of California at Davis, has shown.[57] They lay out how we could provide 100 percent of the world's energy from water, wind, and solar technologies in just two decades—we'd eliminate the use of fossil fuels entirely by the early 2030s. The solar part of their plan would mean building and installing 40,000 photovoltaic power plants (at 300 megawatts [MW] each); 49,000 concentrated solar power plants (300 MW each); and 1,700,000 rooftop units (0.003 MW each). Scaling up wind generation involves manufacturing and putting online 3,800,000 wind turbines (5 MW each). The water-power part of their proposal involves installing 720,000 0.75 MW wave converters in the oceans and building 490,000 tidal turbines (1 MW each), 5,350 geothermal plants (100 MW each), and 900 hydroelectric plants (1,300 MW each).[58]

You may be thinking, "Whoa, those are some really big numbers." But as Jacobson and Delucchi aptly point out, they are not so big when you put them in the context of what people have already demonstrated they can accomplish. Manufacturing and installing 3.8 million wind turbines within thirty years? Entirely doable when you consider that the world manufactures 73 million light cars and trucks every year. Scale up production? Think about World War II, when the United States decided it needed to churn out aircraft fast. Automobile factories were retooled and over the course of about seven years, 300,000 planes rolled off the assembly lines in the United States alone (486,000 more were built in the rest of the world). Deal with changes to infrastructure? It took the United States less than fifty years to go from two-lane roads and relatively little pavement to an interstate highway system that extended for 47,000 miles, enough to encircle Earth twice. Oh, and put a man on the moon? Ten years. Not to mention landing a robot on Mars.

Nobody is saying it will be easy to scale up carbon-neutral energy production quickly enough to solve the climate problem—just that it is possible, given enough commitment and hard work. Additional technological advances will be needed, such as increasing the efficiency of solar cells, developing better power-storage capabilities, and revamping electricity grids to balance power delivery in view of the variable nature of energy from sources like solar and wind. Also, obtaining the raw materials needed to build all those turbines, solar cells, and batteries will be tricky in terms of how it will affect the global economic and political landscapes. For example, Bolivia and Chile will find themselves in the happy position of controlling over half the world's lithium reserves if lithium-ion batteries become even more widely needed. Rare-earth metals are presently essential for turbine gearboxes; China is currently the lowest-cost source. Platinum, which is in short supply, is needed for fuel cells. And of course, critical in the saving-species equation will be extracting those resources in ways that minimize local ecological impacts. Such challenges, while they will certainly complicate things, are not deal breakers, especially considering that researchers are in the process of developing even better ways to generate clean power—like gearless turbines and more efficient batteries.[59]

So, if we have the technical know-how to get rid of the internal combustion engine, produce electric cars, make jet fuel from algae, and replace coal, oil, and natural gas with solar, wind, hydro, tidal, and geothermal power and maybe, just maybe, safer next-generation nuclear plants, why do we keep pouring money into finding ways to get every last drop of fossil fuel out of the ground? The answer is that old ways die hard. Part of it is that sometimes—most times, actually—emotions trump logic. Another part is that it's usually easier to keep doing what you've always done than to strike off on a new adventure. Finally, there is not a compelling *economic* reason to shift from fossil fuels within the time frame needed to solve the climate disruption problem. We're not going to run out of fossil fuel fast enough to make oil, coal, and gas prohibitively expensive even in our children's lifetimes, arguments about peak oil notwithstanding.[60] The situation is pretty much as former Saudi oil minister Sheikh Ahmed Yamani said years ago in regard to what will ultimately push us past the Age of Oil: the Stone Age didn't end because we ran out of stones. It ended because we invented something better.[61]

Now we *have* invented something better, something that can give us the cleaner, sustainable energy we need without the unhappy side effect of killing a good number of species. We just have to scale up production of our inventions and put them to work. That's where getting past social inertia and political gridlock comes in. Human nature being what it is, overcoming the status quo in the short time we've got to make a difference is probably going to require a push from the top, in the form of laws, investment incentives, carbon taxes, and other government policies designed to stimulate the right kinds of innovations and realign the economic landscape.[62] I'm against big government as much as the next guy; but then, I also have to admit that when I was a kid I was against a lot of things my parents told me to do, although in retrospect I see that what they enforced probably kept me alive.

Staying alive: that's really what it's all about, as far as avoiding the Sixth Mass Extinction goes. I still get back to those Colorado hills above Paonia on a fairly regular basis. I've got family down the road, and I also like to go back because I've got lots of good memories of those days I worked in the coal industry. Truth be told, I'm even a little proud when I drive by some of the mines that I like to think went in as a result of our hard work almost forty years ago. What humans can do is incredible. But when I keep on driving up Grand Mesa and get to where I used to see those quaking aspen groves shimmering in the early morning sun, I see they're dead.[63] Whole forests, victims of the climate change that came from burning coal and from spewing all that CO_2 out of my tailpipe, just in my lifetime. That's how I know the past is gone, that the old ways won't work anymore. It's time to get down to business and start building our future.

CHAPTER FIVE

Food

Some of the best fishing I've ever had was on a moonless night, pitch-black except for a few billion stars pinpricking the sky. I couldn't see a thing as I waded into the river, but I had fished that stretch often enough to know exactly where to put my feet as the current pushed against my knees, then my thighs, while I slowly worked my way upstream.

A few hours earlier, I had been hiking into the canyon with my two brothers, as had become our weekend routine that summer. We'd dropped our packs off at the usual camping spot, set up our tents, thrown a couple of steaks on the fire, and topped off the meal with some cowboy coffee. We rigged our fly rods while it was still light enough to see and then made our way down to the water when darkness hit. That night, I got the prime spot, a place we called the rock hole. A house-sized granite boulder had rolled down the steep canyon wall, lodging itself into the far side of the river in just the right position to create an upstream riffle that tailed out into a deep, cold pool that the big trout seemed to love. There was a trick to fishing that riffle, which involved positioning yourself and casting your fly in just the right place, such that the water carried it into the part of the pool where the fish lay low. That night, for some reason, I couldn't miss. My hand-tied nymph would splat upstream just where I willed it to drop, and sink down a bit.

I'd feel the current take it to where I wanted it to go, and then, bam! My rod would bend double, the reel would scream as a big rainbow or brown trout started to run, and we'd battle it out until I slowly, slowly worked the fish in, eventually backing out of the water and coaxing the exhausted trout up onto the bank. There I would admire it in my flashlight's beam, with more than a little anticipation of how good it was going to taste for breakfast in the morning. And then I'd go back into the river for another round.[1]

So I really get it when commercial fishermen say things like "God, I love to catch them. And I know you need some kind of catch limits because I'd catch all of them if I could."[2] That thrill of conquering other species is an almost addictive emotion, and I suspect it is locked deep in our genetic code because it was such a necessary part of life when the only way to get a meal was to catch it—be it a fish, a gemsbok, or anything else. What better way to be successful than to love what you have to do? A similar kind of deep-seated satisfaction comes from transforming a wild, untamed tangle into a productive garden. And when the garden gets really big—farm or ranch sized—it's hard not to feel that all's right with the world when you're out there on the tractor on a summer day, cutting hay that you know will see the stock through the winter, or getting the ground ready for the winter wheat crop.

Nowadays, of course, it's not just feeding ourselves, but feeding the world—producing enough food for the seven billion people currently alive and the two billion or so more on the horizon—that we have to worry about. And therein lies the conflict between food for us and keeping other species on Earth. That conflict arises on two fronts: the first is the amount of land we need to farm and ranch, and the second is the extent to which we rely on nature's bounty of wild species to feed us. On both fronts, other species are beginning to lose.

If you think back to the preceding chapter, you will recall that we—people—co-opt close to a third (about 29 percent) of the land-based net primary productivity that Earth is capable of manufacturing on its own. We have done that largely by transforming 33 percent of Earth's land into farms, ranches, and pastureland. If we consider only ice-free land, the percentage we've used for food production is 38 percent; of that, about 12 percent is cropland and 26 percent is pastureland.[3] Furthermore, we've taken over the best of Earth's lands for our food production. Most of what remains is pretty marginal as far as growing plants and feeding animals goes—we're talking desert, rugged mountains, tundra, and so on. In fact, the productive lands that remain are precisely the ones that would be most problematic to convert to agriculture if the goal is to keep other species alive—forests and grasslands in the tropics and subtropics, which harbor most of the terrestrial biodiversity we have left.

Our need to produce food has both direct and indirect effects on other species. The direct effect is to take that third of the planet's ice-free lands we use for food production out of play for other species. For example, contrast what you would see today with what Meriwether Lewis and William Clark recorded in the years 1804 to 1806 as they traversed present-day Missouri, skirted up through eastern Kansas and Nebraska, and cut north through the center of South and North Dakota and then east across the plains of Montana. Now you mainly see corn, soybeans, cattle, and hogs at the Missouri end of the trek, and not much changes along the way, although by the time you get into the Dakotas and Montana, corn is replaced by wheat, and cows and horses are much more in evidence than hogs. What Lewis and Clark saw instead of monoculture farm fields was a diverse mixture of short and tall prairie grass interspersed with many kinds of shrubs, bushes, and trees. Instead of cows, horses, and hogs as the main megafauna, they encountered bison, deer, elk, pronghorn, wolves, and grizzly bears.[4]

It doesn't take a scientific study to recognize that for most of those formerly abundant and widespread species, the required habitats are simply gone, replaced largely by the low-biodiversity landscapes we have modified to feed us. And wild species that can survive in agricultural habitats typically are not species we want there, so they are killed if they wander in. The classic examples are the big carnivores that sometimes prey on livestock and large herbivores that destroy crops—wolves in the ranchlands of the American West, for instance, and lions or elephants in parts of Africa.[5] Getting rid of such large animals often leads to even more, usually unanticipated biodiversity losses—time and time again, their removal has been shown to decrease the abundance of species and populations in the ecosystems of which they are part.[6]

This may seem counterintuitive at first. After all, don't big predators eat lots of other animals, and don't big herbivores require lots of plant food, so shouldn't their presence decrease biodiversity? In fact, the big predators and herbivores enhance biodiversity by their interactions with other species. For example, in the Greater Yellowstone Ecosystem, the large tract of lightly populated and relatively wild land in the western United States that has Yellowstone Park at its core, wolves keep down the number of coyotes and elk, which in turn increases the number of foxes and allows more willows and aspen to become established. In the case of large herbivores, the right number of elephants (not too many, not too few) promotes a mixed grassland-forest that harbors many more species than ecosystems from which elephants have been removed.[7]

In addition to deleting large swaths of habitat for native species directly, conversion of land to agriculture fragments the remaining habitat. The net impact is that even if species have a suitable place to live on the outskirts of agricultural areas, they have difficulty maintaining viable populations because individuals cannot easily disperse across farm fields and pastures, either because the habitat is unsuitable or because they are apt to find themselves in a rancher's or farmer's gun sight. That means breeding populations stay small, arranged as isolated clusters of a few individuals rather than clusters connected by individuals dispersing from one cluster to the next. Ultimately, isolating those clusters from each other reduces genetic variation and makes populations more susceptible to such unlucky events as a local drought or disease outbreak. All of that means, basically, that in order to drive a species to extinction, you don't have to get rid of all of its habitat. You just have to destroy enough habitat to disconnect and reduce the surviving populations below a critical number.

That critical number is called the minimum viable population, and while it varies considerably depending on the species and time duration of interest, most studies suggest populations of thousands of individuals, not hundreds, are needed if extinction is to be avoided.[8] Those thousands of individuals can be distributed through many discrete and somewhat separated populations as long as those populations are connected enough to allow individuals to travel from one population to the next often enough to promote a certain amount of gene flow. The individual populations, while remaining to some extent isolated, still swap enough genes to stay connected as a larger cohesive group that biologists call a "metapopulation." Today we tend to think of metapopulations in the context of species survival and extinction, and we think of agriculture as interfering with metapopulation dynamics. In that context, it is ironic that the metapopulation concept was formalized in a study that aimed to enhance agricultural production through understanding the population dynamics of insect pests in separate fields.[9]

How conversion of landscapes to agricultural fields acts to reduce population sizes and connectivity is well illustrated by what is happening to Malayan tapirs (*Tapirus indicus*), forest-dwelling mammals in Indonesia. Malayan tapirs look a little like Dr. Seuss invented them out of spare parts. Distant cousins of horses, they are "odd-toed" ungulates, so called because they have three toes (an odd, rather than an even, number) per foot, and the middle toe, capped by a prominent hoof, is the largest. But there the resemblance to horses ends. Moving upward from the hooves, the legs look short and stubby, rather than long and sleek. The body is more piggish than horse-like, the neck is short, the ears look a little like those you would find on an oversized rodent, and the snout is a flexible, stunted, elephant-like trunk. The coloration is also Dr. Seuss–like: in babies, dark gray, with white spots on the legs and cheeks and several white racing stripes running front to back along the body. As they grow up, that striped and speckled coat gives way to the adult garb of a bright white midsection book-ended by formal black from the shoulders forward and the rump backward. That coloration, outlandish as it seems when you see them in a zoo, in fact camouflages them nicely in the forests that the tapirs call home, where interwoven patches of shadow and sunlight pattern the forest floor and thickets.

All too often these days, however, those intricate patchworks are falling to chainsaws as forests are cleared to make way for fields of palm oil plants. Palm oil, long used for cooking in parts of Asia, Africa, and Brazil, has gained ground in the global food industry as a result of the realization that trans fats in cooking oil are not a good thing. Palm oil, which both lacks trans fats and produces good yields of oil per acre, suddenly became lucrative, much to the detriment of Malayan tapirs and many other species that rely on Indonesia's forested habitat to survive. Over the last thirty-odd years, or three tapir generations, populations have dwindled rapidly, falling by 50 percent as palm oil plantations have replaced more and more of the tapirs' forest habitat. As the IUCN puts it, "Remaining populations are isolated in existing protected areas and forest fragments, which are discontinuous and offer little ability for genetic exchange."[10] So long metapopulations, and so long tapirs.

Such direct and indirect impacts of expanding croplands are easy to notice. Less obvious are the effects of the other segment of agricultural land, pasturelands. Pasturelands, recall, far outweigh croplands in acreage, making up the majority of agricultural land use. At first glance, pasturelands often look like wide-open, wild places where native species should be able to thrive, allowing for the fact that the original megafauna now have to share the land with cows. And properly managed, pasturelands can, in fact, serve as valuable biodiversity reserves for threatened species, as visits to my friend's ranch in Montana's Big Hole country used to show me. Walking the ditches with shovels slung over our shoulders (which we needed to plug up gopher and ground squirrel holes that would otherwise breach the irrigation system for the hay fields), we tromped through healthy mixes of sagebrush and grasslands, and along with cows we saw abundant signs of elk, many kinds of small mammals, eagles, hawks, and other birds, and talked about the encounters we'd had on horseback with bears, mountain lions, and moose. And of course, we continued the discussion, always ongoing among Montana ranchers, of whether or not wolves should be hunted out of the area (the ranchers' pretty much unwavering view: yes, as fast as possible).

Not much of the 26 percent of Earth's lands that have been converted to pasture look like that idyllic ranch in the Big Hole. Instead, in most places, livestock grazing tends to irretrievably change the ecological underpinnings of pasturelands. Some of the changes are obvious even to the casual observer, for example, overgrazing that results in soil degradation and loss of productivity.[11] In extreme cases, that turns marginally productive grazing land into barren desert, as happened in the infamous example of the Sahel region of Africa. In that instance, the combination of overgrazing and deforestation, added to a few years of drier-than-normal climate, caused widespread starvation for people in the region and a permanent loss of arable land.[12] That same combination—climate change plus overgrazing or other means of destroying vegetation—in fact has been implicated as a general driver of desertification, which is of particular concern given the climate-change issues I discussed in the previous chapter.[13]

Such extreme impacts, thankfully, are not very common. Instead, more subtle pasture modifications markedly reduce the habitat available for non-domestic species, and even for the domestic species the pasturelands are supposed to support. In the vast tracts of grazing land in the American West, for example, it is not overgrazing that is the biggest problem (although that certainly occurs, and not uncommonly), but takeover of the native, palatable grasses by nastier immigrants like cheatgrass, technically known as *Bromus tectorum*. Cheatgrass is an invasive species. It arrived in the United States via contaminated crop seed and ship ballast sometime in the late 1800s, from its original home in southwest Asia. It has spread throughout the American West, with one of the main vectors being livestock.[14]

That spread, slow at first, has been accelerating. Over the past three decades cheatgrass has reached epic proportions, and the results are not pretty. In fact, cheatgrass has been known to make me curse, as my wife, Liz, can attest. Most summers our field research takes us into seemingly endless, isolated pasturelands in places like Montana, Wyoming, Idaho, Oregon, Nevada, Utah, and Colorado. We and our students end up doing a lot of hiking through areas where cattle are grazed, such as on lands administered by the Bureau of Land Management and the Forest Service, and through areas where grazing is minimal or absent, such as on National Monument or National Park lands. Thirty years ago on those long walks, no matter which agency managed the land, we usually scrambled through a variety of different grasses interspersed with sagebrush. What we began to notice over the past ten years, though, is that walking through those places is now a very different experience—cheatgrass growth has exploded such that it is the dominant grass, and it doesn't take long for our boots and socks to fill up with its perniciously burrowing, sharp, and immensely irritating seeds.

I've been told that Oregon ranchers claim the name "cheatgrass" derives from the fact that it was once common practice to cheat ranchers who were buying hay for the winter by bulking up the bales with the fast-growing but not particularly nutritious *Bromus tectorum*.[15] The chemical company BASF, in a brochure advertising chemicals used to kill the plant, claims the name comes from the effect the grass has of "cheat[ing] landowners and ranchers from earning the full economic benefit of their land by displacing native plants, reducing biodiversity and spreading fires."[16] All of which cheatgrass does.[17]

As a result, cheatgrass is not a particularly welcome plant in the American West. The problems are many. It is a prolific seed producer and outcompetes native grasses in short order. It does particularly well in disturbed soils, that is, soils that are trampled by livestock, or along road or railroad right-of-ways. It is spread rapidly by the livestock that eat it, which deposit the seeds in a nice blob of fertilizer: cow dung. Although eaten by livestock, it's not nearly as good for them as the native grasses. While it is palatable before the seeds form, after seeds appear the grass's nutritional value decreases markedly, and the sharp seeds injure the animals by burrowing into their mouths and eyes and causing infections. Adding insult to injury, cheatgrass ends up competing with and reducing yields of decent forage like alfalfa.[18]

Not to mention cheatgrass's propensity to reduce biodiversity where it begins to dominate. This is not a particularly subtle effect, if you know what to look for. Just a couple of years ago, Liz and I and several of our students set out on an expedition to catalogue small-mammal biodiversity along a transect through California, Nevada, Utah, and Colorado. We collected data from two sources to figure out what small mammals were common. One source was the regurgitated pellets of owls, hawks, and other raptors, which contain bones of the small mammals, reptiles, and amphibians that the birds eat. The raptors generally do a better job of sampling what small animals are out there than a mammalogist does with his or her traps; all you have to do is separate the bones from the nasty stuff, do a few statistics, and you end up with a pretty good census. The other data source we were after was rabbit poop, which, after Liz and her lab group extracted the DNA from it, could reveal which species of rabbit inhabit the region. What quickly became apparent, as we trudged around under the high-desert sun, was that the more cheatgrass there was, the fewer raptor pellets we found, even when adequate raptor roosts were present. This implied fewer small mammals on the landscape and, consequently, fewer big birds that rely on eating small mammals. Likewise, the frequency of rabbit poop on the ground diminished with increasing cheatgrass, to the extent that in areas where cheatgrass was rampant, there was virtually no sign of rabbits. Those qualitative observations were borne out when we got back to civilization and began poring over the scientific journals. It turns out that recent studies document that the denser the cover of cheatgrass, the less abundant the small mammals, both in numbers of species and numbers of individuals. That's particularly true for species that specialize in eating seeds, which, in the dry parts of the American

West where cheatgrass grows best, happen to be most of the rodent species.[19]

You begin to get a sense of how important such direct and indirect habitat transformations are in the extinction equation, and how much agriculture contributes to those, when you scale up to the global level. Available numbers indicate that habitat destruction is the chief threat for 86 percent of all birds, 86 percent of all mammals, and 88 percent of all amphibians presently threatened with extinction.[20] And the number one cause of habitat destruction is—you guessed it—conversion of forests and grasslands to farms and pastures. In fact, a study from 2002 implicates agricultural expansion in 96 percent of all instances of tropical deforestation.[21]

Those stark facts make it pretty clear that avoiding massive extinctions of other species in the coming years will require halting expansion of agricultural land. That is a tall order when you consider how many more people we will need to feed by 2050. Presently, the acres of land we devote to agriculture (crops plus pasture) divided by the number of people in the world (about seven billion) indicate that we're using about 1.7 acres per person for food production. Let's say we continue agricultural expansion at that ratio to keep pace with the growing number of people on Earth, which is estimated to reach 9.5 billion by 2050 (remember, most demographic models say we are destined to hit at least that number, but it could be a lot higher or a little lower, depending on what happens to birth rates, especially in poor countries, over the next thirty years). At 9.5 billion people, each requiring their 1.7 acres to obtain food, we would need to devote about 50 percent of Earth's ice-free land, rather than the present 38 percent, to feeding people. Converting that additional 12 percent to agriculture would mean, quite simply, that rainforests are doomed. We'd have to cut down every rainforest on the planet, because they are among the few arable lands that we haven't yet changed into farms or pastures. That would basically put the nail in the coffin for bringing on the Sixth Mass Extinction, given that rainforests support most of the terrestrial species in the world. But converting all rainforests to agriculture would get us only 5 percent of the additional 12 percent we'd need, so then we'd have to figure out a way to farm really inhospitable places, like deserts, rugged mountains, and tundra.[22] Good luck with that. Clearly, we cannot keep on using 1.7 acres per person for food production. We're going to have to get much more efficient than that.

Fortunately, all indications are that we can. First, our efficiency, in terms of acres per person, has been increasing since 1950, and especially since the Green Revolution went into full swing in the late 1960s. That, as its name implies, changed the very foundations of agriculture, shifting it from simple subsistence to big business. The breakthroughs that resulted in marked increases in food yield per acre included development of better seed stocks, expanded irrigation, introduction of synthetic fertilizers and pesticides, and a shift from small family farms to large, mechanized factory farms complete with sophisticated management structures. While the societal, cultural, and environmental impacts (especially the loss of family farms and the increased use of pesticides and fertilizers) in many cases left much to be desired, there is no arguing with one fact. Per acre, more people got fed, and at a lower cost. From 1950 to 2012, acres of agricultural land increased only a minuscule amount, while world population grew from about 2.5 billion to 7 billion; the acres per person used for agriculture fell from about 4.8 to the present 1.7, more than a doubling of efficiency.[23]

Recent work by a number of food-security researchers suggests that this trend can continue. It is well within humanity's grasp to avert the extinction costs that would result from desperately converting more and more land to agriculture—an approach that would be doomed to failure in any case, because there simply isn't enough arable land to solve the problem by expansion. The much more feasible alternative lies in four coordinated actions: closing the "yield gap," being more efficient in how we produce food, eating less meat, and wasting less food.[24]

An article published in the journal *Nature* in 2011 titled "Solutions for a Cultivated Planet" summarizes key parts of most of these approaches.[25]

Closing the yield gap refers to increasing crop yield on lands presently in agricultural production but not operating at full capacity, especially the world's least productive farms.[26] Just closing that gap for the world's most utilized sixteen crops would result in a 50 to 60 percent increase in food production. Doing that will require refining crop strains, through both traditional breeding methods and genetic modification, to ensure that plants are maximally adapted to local growing conditions.[27] Very important will be maintaining a diversity of strains; we need to have seed stocks available that will allow us to respond quickly to what promise to be rapidly changing climatic conditions over the next few decades, when many crop-growing regions will undergo shifts in patterns of temperature and rainfall.[28]

Of course, closing the yield gap does not do much good, from the perspective of extinctions, if it introduces more environmental problems than it solves. That's where the second part of the solution comes in: using the resources required for food production much more efficiently, by starting with better seeds, but especially by restricting fertilizer and pesticide use, employing efficient irrigation, and using planting techniques that minimize soil disturbance and erosion. Amazingly, those "simple" fixes can result in pumping up production dramatically without causing additional environmental damage.

The gains that can be made are truly staggering. For instance, using drip irrigation, rather than spraying water into the air; mulching; and designing canal and reservoir systems to minimize evaporation can reduce the present use of about one liter (or quart) of water per calorie of food grown by as much as 70 percent.[29] Water conservation issues are not trivial considering that 70 percent of all water currently withdrawn from rivers, lakes, and underground aquifers is used for agriculture.[30] "Water wars" are already a major source of contention among urban dwellers, farmers, and environmentalists, and that situation promises to worsen as climate disruption causes precipitation patterns to shift at the same time that demand for water increases, not only for agriculture but also for energy production and to meet the needs of cities.[31] While farm-by-farm water-conservation measures will certainly help, accruing maximum water savings and profits for farmers will require a basin-wide approach tailored to maximizing water conservation over entire watersheds, combined with incentives to implement water-saving technologies, such as putting the right price on water.[32]

Fertilizers are another area where the gains can be spectacular.[33] Fertilizers are currently applied in quantities that far exceed their need, particularly in certain parts of the world (the western United States, China, northern India, and western Europe). In fact, much of the pollution from fertilizers—30 to 40 percent—comes from only 10 percent of the world's croplands. By applying fertilizers with what has been called the "Goldilocks strategy"—not too much, not too little, but just the right amount—crop yields can be increased and fertilizer-generated pollution decreased.[34] The Goldilocks strategy can be greatly enhanced when local, state, and/or federal policies reward farmers for meeting agreed-upon standards for limiting nutrient runoff.

I get a little wistful thinking about the third fix to the agricultural system that is required to keep other species alive. I'm a meat eater. My dad was a butcher. There's nothing like a good prime rib. But the numbers don't lie. Nearly a quarter of all croplands go toward feeding animals. And—brace yourself—75 percent of *all* agricultural land (that's cropland plus pastureland) is used for feeding livestock. Remember back to the preceding chapter, where I talked about the energy lost as you move from one trophic level (for example, cows) to the next one (us)? If we divert that caloric energy to ourselves by eating the crops we grow, rather than first feeding the crops to livestock that we subsequently eat, we increase the world's available caloric yield by about 23 percent and food production by about 13 percent.[35] Does that mean we have to quit eating meat altogether? From my perspective, the answer is thankfully no. Raising cows to eat is a perfectly good use of most of the pastureland presently in production, as long as the right numbers of cows are run per acre. Much of that pastureland is in arid or otherwise poor crop-growing regions, so as long as it is managed properly, the best agricultural use of it is meat production. What we have to do away with are feedlots that fatten up cows with corn and other feed crops that take up so much of our prime cropland. That means less meat, for sure, but it also means less fatty, healthier, and (in my opinion, at least) better-tasting meat.

It's not just using prime cropland for raising animal feed that's cutting into maximum food production, though. The other big chunk of lost calories is biofuels. Crops devoted to biofuel production now cover about 4 million square kilometers (roughly 988 million acres) worldwide, or about 26 percent of total cropland area. Devoting that land to food production would increase global food supplies by nearly 15 percent. Here is where, at first glance, a conflict arises between moving away from fossil fuels and feeding the world. In fact, that conflict largely goes away if biofuel production from sources like algae ramps up instead of biofuel produced from crops grown on prime food-growing lands.

Another huge positive input to the food equation is to simply be less wasteful.[36] Somewhere between a third and a half of all food grown never makes it to people's mouths because it spoils or is otherwise rendered inedible in the journey from farm to processing to storage to delivery. And in industrialized countries, once food is delivered to the consumer, 40 percent of what is served on our plates ends up in the trash can.[37]

Implementing these agricultural solutions would more than solve the dilemma of how to keep other species alive and still feed a human population that will increase by about 26 percent, all without cutting into lands that aren't already in agricultural production. We can get 40 percent more food by closing the yield gap, 28 percent more by growing food for human consumption on lands now used for animal feed and biofuels, and another 30 to 40 percent by reducing food waste. These solutions do not even take into account gains from such innovative approaches as installing rooftop gardens on new buildings, replacing part of your yard with a vegetable garden, and, in cases where it makes sense, growing crops in a mosaic that includes native vegetation

(I'll talk about this a little more in the next chapter). All of these solutions are eminently doable technologically, but it will take the willingness of individual farmers and ranchers, the food industry, the general public, and government to prioritize the goal of feeding the world without expanding the agricultural footprint.[38] It will also take starting today: we have only thirty-five years to ramp up food production to the extent needed to feed those extra 2.5 billion people who are on the way.

Growing our food is not the only issue we have to worry about if we want to avoid the Sixth Mass Extinction—we also have a tendency to overexploit wild species to put food on the table, to the extent that we're all too effective at hunting the biggest, tastiest wild animals to extinction. That first became evident ten thousand or so years ago, as the Pleistocene epoch was giving way to the Holocene, when the big delicious animals were mammoths, mastodons, giant ground sloths, and other like-sized beasts.[39] By the twenty-first century, we've moved on to smaller-bodied species. Most at risk from hunting are species that are utilized in the bush-meat trade, particularly in parts of Africa.[40] These include animals like gorillas (*Gorilla gorilla*), chimpanzees (*Pan troglodytes*), and forest elephants (*Loxodonta africana*), some of which are hunted for meat, others of which are accidently caught in snares or otherwise killed in pursuit of the meat species. A key difficulty is that bush-meat species often provide a much-needed source of protein for local people.[41] To get around that dilemma, numerous efforts are under way to make other sources of protein (for example, chicken or fish) readily available. A second problem involves the income that can be earned from selling bush-meat and other wildlife. I'll go into economics in a bit more detail in the next chapter, but for now, suffice it to say that the solutions to the economics of using bush-meat for food lie in some combination of the following actions: addressing local population growth, such that food shortages do not require hunting local wildlife; developing livelihoods that do not rely on hunting; better land-use planning; enforcement of hunting regulations; wildlife ranching; and making available international funding to help local communities overcome the obstacles inherent in switching from bush-meat to more sustainable food sources. All of these approaches are being experimented with to varying degrees.[42]

As important as our impacts on land are, there is also the very big problem of what our quest for food is doing to the oceans' species. One food-related threat to sea-dwelling species results directly from agriculture on land: dead zones that spread along coastlines as a result of agricultural pollutants, especially fertilizer. Applying too much fertilizer (and animal waste from feedlots) results in excess nitrogen runoff, which now elevates nitrogen levels in the waters off virtually every coastline in the world. The dead zones that result are enormous. A single one in the Gulf of Mexico is at least the size of the state of New Jersey.[43] As of 2008, dead zones, which are essentially devoid of marine life, had spread through about 95,000 square miles (245,000 square kilometers) of ocean, an area roughly equivalent to the entire United Kingdom.[44]

Widespread habitat destruction and fragmentation also exist at sea. The chief culprits are bottom trawls (essentially, weighted nets) that scrape the ocean floor, leaving a largely barren seascape. This has the same effect as the chainsaws and plows we use on land: the area available for nonhuman species to live on gets smaller and smaller. Bottom trawling has been estimated to disrupt an area of the sea floor twice the size of the continental United States—each year.[45]

But the biggest immediate problem that the oceans' species face is over-fishing, which is not so different from the over-hunting that helped trigger the end-Pleistocene megafauna extinctions I mentioned earlier. In fact, it is probably no coincidence that the biggest oceangoing species—certain species of whales—were the first marine animals to be hunted nearly to extinction in the 1700s and 1800s. As a result, blue whales (*Balaenoptera musculus*) and right whales (*Eubalaena glacialis, E. japonica,* and *E. australis*), along with several other big-bodied whale species, are hanging on by a thread.[46]

Today, in addition to whales, we are over-exploiting fish. Whereas much whale hunting was to obtain whale oil and other non-food products, over-fishing is almost entirely driven by our desire to satiate our palates. Among the species we are fishing to death are bluefin tuna (*Thunnus maccoyi, T. orientalis,* and *T. thynnus*), Atlantic cod (*Gadus morhua*), beluga sturgeon (*Huso huso*), and various species of salmon and sharks. We are catching and eating these fish faster than they can reproduce, because, just like those woolly mammoths way back when, they were present in seemingly indestructible numbers in our fathers' time, and old habits (and acquired tastes) die hard. The difference between our fathers' time and now, of course, is that the human population has more than tripled in the last two generations. It isn't too hard to see the numbers problem—there are now seven times as many humans on the planet as there are wild salmon.[47] There simply aren't enough wild fish in the sea—at least, of the species we're fond of eating—to keep feeding us at the rate we've been using them.

Going after those wild fish species also results in "bycatch," that is, scooping up unintended species along with the fish that are destined for our tables. This is not a trivial problem as far as extinction threats are concerned. As of 2012, bycatch was the main threat for marine mammal species that are in the IUCN's vulnerable, endangered, and critically endangered categories.[48] This includes the northern fur seal (*Callorhinus ursinus*), the Australian sea lion (*Neophoca cinerea*), and the Irawaddy dolphin (*Orcaella brevirostris*).[49] Likewise, becoming entangled in fishing nets or on hooks is the principal cause of death for threatened marine turtles. One study documented that in the year 2000 alone, 200,000 endangered loggerhead turtles (*Caretta caretta*) and 50,000 critically endangered leatherbacks (*Dermochelys coriacea*) ended up on longline hooks.[50]

It's easier than it ought to be for us to fish species to extinction for a couple of reasons. First, people are a different kind of predator than many animals that end up at the top of the food chain. When we run out of one species, we simply move on to another, in contrast to the many predators that tend to focus on a single or small group of related species. Predators that specialize in only certain species in fact ensure that they don't cause their prey species to go extinct, because the survival of the predator species depends on a long-term balance between them and their prey. If the predators eat too many of their prey, the next generation of prey is not very abundant, and there is not enough for the predators to eat. That results in a subsequent crash of the predator population; then the prey population recovers, which subsequently supports growth in the predator population. Such regulation of both predator and prey populations has been observed for Canadian lynx and their snowshoe hare prey.[51]

In contrast, we people, being adaptable in what we choose to prey upon and the tastes we develop, tend to smack our lips over the last morsel of a species about to go extinct as we set our sights on the next species in line. This sort of adaptability seems to be precisely what allowed relatively small populations of prehistoric humans to cause such carnage in big-animal populations, resulting in the extinction of about half of the big-bodied mammal species on Earth near the end of the Pleistocene.[52]

The other reason it's easy for us to over-exploit fish species is that we tend to think in terms of *kinds* of fish, rather than *species* of fish. Tuna is a prime example, both of this principle and of moving from one prey species to the next. There are actually eight species of tuna, not just one. Of those, the most highly prized ones (they make great sushi) are commonly lumped under the single moniker bluefin tuna, a name that actually encompasses three species: the southern bluefin (*Thunnus maccoyii*), the Atlantic bluefin (*Thunnus thynnus*), and the Pacific bluefin (*Thunnus orientalis*). First in line for fishing were the southern bluefins, maybe because they were tastiest, maybe because they fetched higher prices, maybe because they were just easier to catch. Whatever the reason, the species started to be intensively fished in the 1950s. Because southern bluefins also have a long generation time and the biggest fish (the ones that yield the most dollars per effort) are the oldest, it didn't take too many decades to pull a good number of the fish that were in their reproductive prime out of the spawning stock. As a result, between 1973 and 2009, 85 percent of the spawning biomass (fish of spawning age) perished, and currently adult biomass is estimated to be 5 percent of what it was prior to heavy fishing pressure.[53] No wonder, then, that the fishery has crashed, with no evidence that it's coming back. Neither will the species come back without intensive conservation efforts; *Thunnus maccoyi* is now considered critically endangered.

As the southern bluefin stocks were being depleted, Atlantic bluefins began to take a harder hit. Beginning in the 1960s, fishing pressure on them intensified, with their spawning biomass dropping by 51 percent from the 1970s to 2010, placing them in the endangered category.[54] They continue to be fished far above sustainable yields. If that keeps up, it is only a matter of time before the Atlantic bluefin fishery crashes, just like the southern bluefin fishery did.

Pacific bluefins are next in line. You might think there is nothing to worry about if you run across them at your local sushi restaurant and if you bother to check the IUCN Red List. The 2011 assessment identified them as a species of least concern, noting that catch numbers were relatively stable from the mid-1980s to 2011. But the IUCN assessment came out before a 2013 study that pointed out that Pacific bluefin numbers have dropped 96.4 percent relative to pre-fishing levels, and more than 90 percent of the Pacific bluefin caught are juveniles, which means they are eaten before they have a chance to reproduce.[55] Currently there are no catch limits for fishing them in the western Pacific. If that is not a recipe for extinction, I don't know what is.

Closely related to bluefin tuna are two other species, bigeye tuna (*Thunnus obesus*) and albacore tuna (*Thunnus alalunga*). Bigeye tuna have seen a 42 percent decline in their spawning biomass in the past fifteen years, and albacore a 37 percent decline in theirs, rendering the two species vulnerable to extinction and near threatened, respectively.[56]

At least two of the other three species of tuna are, like the bluefin, bigeye, and albacore, clearly on a downhill slide relative to their pre-fishing numbers. Yellowfin tuna (*Thunnes albacares,* also known as ahi tuna) is the one I see most commonly as steaks in stores these days. Fished from tropical and subtropical seas through most of the world, yellowfin "only" lost an estimated 33 percent of their global population over the years 1998–2008.[57] Until recently, the IUCN classified yellowfin as a species of least concern, but because of increased catches by commercial fishermen in the Indian Ocean, the species now is considered to be in the near threatened category. Blackfin tuna (*Thunnus atlanticus*), designated as a species of least concern by the IUCN, seem to be holding steady so far. The eighth tuna species is the longtail tuna (*Thunnus tonggol*), about which so little is known that it's impossible to say whether or not it is doing well.

If we continue on the trajectory of the past three decades, at least 75 percent of tuna species will be off of our tables and likely off of the planet within this century. Now magnify that by all the other kinds of fish we're pushing toward extinction. Only a small percentage of fish species have actually been assessed by the IUCN, but of those that have, 1,631 species (28 percent) of bony fish (actinopterygians) and 177 species (17 percent) of sharks (chondrycthians) are now threatened with extinction, mostly because we like to eat them.

There are, fortunately, solutions at hand that can help defuse this part of the extinction crisis. One thing that can help—if done in the right way—is aquaculture, or fish farming. However, that caveat of doing it right is a big one. Aquaculture is already a huge industry—hundreds of species of fish and shellfish are farmed to the tune of billions of dollars and millions of tons of fish per year.[58] Fish farms, in fact, supply almost all of some fish, like Atlantic salmon and tilapia, that you see at the fish counter and in restaurants. Those two fish in many respects represent the good (tilapia) and the bad (Atlantic salmon) that fish farming can do.[59]

Let's start with the bad. In the mid-twentieth century, farming Atlantic salmon (*Salmo salar*) seemed like an environmentally sound thing to do. The species was dwindling in the wild, there weren't enough of them in wild fisheries to feed our growing appetites, and it turned out they were fairly easy to farm in submerged cages. But things soon got out of hand. Those salmon cages have now been emplaced in waters off many parts of the Atlantic coastline, as well as in places that had never seen an Atlantic salmon, like the Pacific coast of the northwestern United States and southern Chile.[60] The effects have turned out to be pretty much the aquatic equivalent of those feedlots for cattle you can smell coming for several miles when you are driving through certain parts of, say, California or Kansas. A single salmon farm containing 200,000 fish (farms often contain four to five times that many salmon) produces fecal matter equivalent to the untreated sewage of 65,000 human beings. The result: dead zones around the salmon farms.[61]

The pollution problem is by no means the worst impact: other problems with the big Atlantic salmon farms include transmission of disease to wild stocks (notably sea lice and infectious salmon anemia, or ISA); escape of farm fish, which interbreed with and weaken genetic stocks of wild salmon; in Pacific waters, introduction of a non-native species that competes with native salmon; and contamination of surrounding waters with huge doses of antibiotics.

However, from the point of view of contributing to the extinction of other species, the worst problem is the fact that salmon are carnivores (technically, piscivores). To keep them healthy, you generally have to feed them some other kind of fish. It takes three pounds of fish meal—generally produced by grinding up wild herring, anchovies, and sardines—to produce one pound of salmon flesh. That means for every pound of salmon raised in a cage, about three pounds of wild fish are lost from the sea.[62] Multiply that by a few hundred thousand salmon in a single fish farm, times many fish farms, and the number for ground-up wild fish gets very, very large. The overall effect is to actually reduce the availability of some kinds of seafood for humans, as well as to reduce the food supply for wild fish-eating fish and seabirds.[63]

This is not to say that all salmon farming is bad—in fact, when done right, salmon farming can be a solution rather than a problem. Rosamond Naylor, a Stanford University food-security researcher who has extensively studied the pros and cons of aquaculture in general and salmon farming in particular, suggests that measures like forgoing antibiotics, using low-density pens that are fallowed and rotated, substituting organic formulated feeds with non-fish ingredients for fish meal derived from wild fish, and regularly testing for water quality could make salmon farming ecologically sustainable.[64] But Naylor is also a realist—given that countries that have invested huge amounts of capital in salmon farming aren't going to be enthusiastic about passing the laws it would take to regulate the industry properly, she thinks the best bet could be for major food chains and other distributors to label sustainably farmed salmon with a "green label" so consumers know what they are getting.[65] Such certification programs, developed by several different environmental oversight groups and some governments, are already in place but need to be refined before uniform, meaningful standards are routinely reflected on marketed products.[66]

By any measure, at the top of the "green list" would be a new breed of salmon farm that is bringing back a species whose populations have crashed dramatically—Chinook salmon (*Oncorhynchus tshawytscha*).[67] This promising approach to salmon aquaculture uses human-made agricultural habitats to mimic the salmon's disappearing wild habitat and preserve the salmon's in-the-wild life cycle, rather than adapting a wild species to life in a cage. Chinooks were once common in California rivers and off the coast but became increasingly scarce over the past century, thanks in part to the draining of floodplains along the Sacramento River. Now juvenile chinooks are being reared in rice fields that are flooded between planting seasons, where the fish are growing longer and fatter and in general seem in even better shape than their counterparts in natural rivers and hatcheries.

Also at the "good" end of the range of aquaculture ventures is tilapia farming, a venture that has scaled up dramatically over the past few years. Tilapia farms provided about 475 million pounds of fish to North Americans in 2010, a fourfold growth over a decade.[68] Tilapia are part of a diverse group of fishes called cichlids that are native mostly to Africa. They include several species in a genus that, like the common name, is also labeled *Tilapia*. Compared to salmon, they offer a huge advantage in quenching our appetite for fish because they are the fish version of vegetarians, which means that you don't have to put more fish flesh into the farm than you can get out of it. Tilapia is also the ideal farm fish—it grows rapidly, does not mind crowded conditions, and, importantly, is a freshwater fish, so it can be raised in tanks and confined to lakes in restricted watersheds or, even better, raised in ponds isolated from natural waterways. Such characteristics mean that farmed tilapia (and other fish like them) can contribute much to providing a tasty source of fish protein while solving the problem of taking too many fish out of the sea.[69] Of course, tilapia farms are not immune to the sorts of problems that affect salmon farms—overcrowding that leads to pollution, escaped fish, and so on—but those problems do appear to be more tractable to deal with in tilapia and similar fish, given their biological characteristics. As with salmon, however, oversight of tilapia farming, cooperation between private enterprise and government, and consumer awareness stimulated by certification of farms with best practices will be required for a viable future.[70]

While fish farms may reduce the demand for some wild fish, they will probably never eliminate it. The only way to deal with that reality is to regulate the number of fish caught by establishing and enforcing sustainable catch limits. Sustainable catch limits must have two provisos: first, to ensure that not too many fish are caught; second, to ensure that of the fish caught, not too many are below reproductive age. The second proviso is just as important as the first, because as a fishery is depleted, the caught fish are younger and younger, until most of the fish end up on our dinner plates before they have a chance to reproduce.

In fact, the needs for imposing reasonable catch limits have long been recognized by the organizations that try to regulate fisheries. Typically, those groups include fisheries biologists, policy makers, and the people who do the fishing. Figuring out appropriate limits involves using the catch records from a number of years to assess whether fewer and fewer fish of a given species are being caught each year. If catches are declining, the trick is to adjust limits to levels that will allow populations to rebound to viable levels and then become stable. The catch limit that would, over the years, sustain populations without danger of collapse is called the maximum sustained yield. Setting and enforcing the maximum sustained yield has proven both feasible and effective. For example, whereas historically, Atlantic bluefin tuna were exploited at numbers triple what would be sustainable, stricter monitoring and compliance measures have resulted in catch reductions of nearly 75 percent in recent years, allowing populations to begin to rebuild. There are similar success stories for North Atlantic swordfish and Spanish mackerels.[71]

Anything that smacks of regulation, of course, immediately raises hackles in many circles and can get you into a good fight at the dinner table. But in the case of fisheries, it's more than rules for rules' sake—catch regulations are the only hope for saving a way of life that goes right back to the little story with which I opened this chapter. Fishermen and fisherwomen love to fish. Just as that river I used to fish would be empty of trout if it were legal for everybody to take home as many as they could catch, so too would commercial fisheries in the ocean collapse one after the next if there were no catch limits enforced. That is something that the sport fishing industry and sport fishers themselves have long recognized and embraced in order to keep the good fishing coming. And it works. I recently returned to the river I used to pull big trout out of so regularly, and the lunkers are still hanging out in the same riffles, in all their glory. What keeps them there, even after four decades of heavy fishing, is strict state regulation governing what you can use to catch the fish, how many you can catch, and a strong catch-and-release ethic on the part of the enthusiasts who fish those waters. In the final analysis, it's the fishers themselves, not the state game wardens, who are most effective in enforcing the limits. Believe me, you don't want to be the guy that the hard-core fly fisher on that river spots using an illegal live worm for bait, or walking down the canyon with an oversized load of fish on your stringer.

A similar self-regulating, self-policing model is beginning to take hold in commercial fisheries, at least in some places. The new breed of regulatory boards employs the commercial fishers themselves to help develop the catch limits—who better knows how dramatically their take is decreasing year to year?—and to help police the waters to make sure others are not taking more than their allotted share. That brought the lobster fishery off the coast of Maine back from the brink of disaster. The key to such win-win solutions is that they focus not so much on saving fish for the fish's sake, but on saving the fishers' way of life.[72]

Consumers also play an important role in regulation. If you don't buy a particular fish at the fish market, the store loses money, which means it quits stocking that species, which in turn means there's not much point in fishing for it commercially. Currently, there are three obstacles that need to be overcome if consumers are going to be effective in helping to guard against unsustainable fishing. The first is that it's hard to know whether a fish you are buying for dinner is one that is doing fine or one that is on an extinction trajectory. Even paying close attention to IUCN criteria is sometimes confusing, since the assessments don't happen at regular enough intervals to reflect recent population crashes, as exemplified by the Pacific bluefin tuna mentioned above. Useful, convenient aids do exist, however—for example, the Seafood Watch effort run by the Monterey Bay Aquarium, which provides easy-to-use cheat sheets you can get as an app for your phone or as a wallet-sized guide.[73] Such guides often provide a more updated status than is possible for the IUCN and tell you not only what kinds of fish are sustainable as food but also what questions to ask your fishmonger: Where's that cod from? (Iceland and northeast Arctic, ok; off the U.S. and Canadian coast, don't buy.) How was it caught? (Hook and line, good; trawling, bad.) Is that salmon farmed or wild? (Avoid farmed Atlantic salmon; wild Alaska salmon are the best choice.) What species of tuna is that? (Southern bluefin, forget it; albacore, I'll take it as long as it's caught by trolling or pole and line and is from the northern Pacific.) If your fish salesman doesn't know, don't buy it. It won't take too many customers asking the same questions and walking away without buying before the owners of the store learn the answers and make sure they are buying sustainable, rather than threatened, species.

The second obstacle is a tougher one. Other than taking the word of the guy behind the counter, there is presently no easy way to know for sure what you are buying. And to be fair, the guy behind the counter, or your waiter at the sushi restaurant, may not know either, because there are seldom regulations in place that require labeling fish by species rather than by "kind," let alone standards that would require labeling the viability of the fish stock from which the fillet originated. Tuna again provide a great example. In the United States, for instance, the federal Food and Drug Administration is the agency charged with identifying, certifying, and labeling food for customers. Its standards do not differentiate between the eight different species of tuna—all eight species can be sold simply with the identifier "tuna."[74] The problem that results is well illustrated by a recent study that used DNA barcoding to identify the species of tuna served in sushi restaurants. DNA barcoding involves taking a sample of the sushi and using molecular biology techniques to determine a small part of its genetic code. That genetic information is then used to match the piece of sushi to the species from which it was cut.

The results: "Nineteen of 31 restaurants erroneously described or failed to identify the sushi they sold.... Twenty-two of 68 samples were sold as species that were contradicted by molecular identification."[75] Oops. Besides contradicting the menu's (or waiter's or chef's) representation of what kind of tuna was being served, in some cases barcoding demonstrated that what was being served as "white tuna" in fact was snake mackerel (also known as escolar, or by its scientific name, *Lepidocybium flavobrunneum*), a fish that has been banned for sale in Japan and Italy because it can cause gastrointestinal problems. The same study went on to point out how nice it would be if everyone had their own cell-phone-sized DNA tester to poke into the fish they were buying and read out what species they were really going to eat. That technology does not yet exist but may someday. In the meantime, a solution that might work almost as well is for food-inspection agencies to require species-level identifications for fish sold in stores and restaurants, with prominent signage that indicates the threat status of the species. Inspection agencies can use existing technology to routinely and randomly genetically test the species being sold, much as health inspectors drop into food establishments to assess concerns.

The third consumer-based issue is probably the trickiest: what we are willing to pay to get what we want. One effect of certain species becoming scarcer is that their price gets higher at the supermarket. The last time I bought a yellowfin tuna steak, which, recall, is a species that recently slipped from being a species of least concern to near threatened by IUCN standards, I paid nearly $30 per pound. And that's nothing. A single 489-pound Pacific bluefin tuna—this is the species that recent analyses have shown has lost 96.4 percent of its population—recently sold in Tokyo for $1.76 million, or about $3,600 per pound.[76] That outrageous price for a fish highlights the third big thing we have to deal with if we are to avoid causing the Sixth Mass Extinction—money, with a capital *M*.

CHAPTER SIX

Money

I count myself extremely lucky never to have been shot at by an AK-47. So I can only imagine the scenes that took place in the Democratic Republic of the Congo in 2012 and that these days are taking place all too often across Africa.[1] I envision the park rangers with their heads down, bellies pressed hard into the tall grass, a little panicked as bullets whizzed through the air, a helicopter off in the distance reverberating behind the staccato gunfire. Hard to say what they were thinking, but my guess is it was something along the lines of "Shit! Out-manned, out-gunned, and those damn bandits just killed another herd of elephants!"

To the bandits and the people they worked for, those elephants were ecosystem services, the fastest kind of natural capital. Biologists, economists, and, increasingly, the world at large recognize ecosystem services as the direct benefits people get from the natural world around them, and the species and ecosystems around us as the natural capital that provides those services.[2] For the elephant poachers, the ecosystem services are theirs for the taking, first come, first served. Hack off the tusks, get them to China, and the drug lords or terrorist groups running the operation pocket a few million dollars. The bandits on the ground manning the AK-47s, and those in the helicopter providing aerial support, may wind up with a couple of hundred dollars—still big money for rural Africa.

That's not the only way elephants provide ecosystem services and natural capital, of course. Shift the scene to Amboseli National Park in Kenya. Ivory raiders hit there too, on a not-infrequent basis, but this morning, that's not foremost in anyone's mind. A little before dawn, we wake up, stumble out of our cabins, and grab a quick cup of coffee. Alpenglow is turning Mount Kilimanjaro, looming just across the border in Tanzania, pink as we pile into open Land Rovers and bounce down a two-rut road in the morning chill. Through the mist, off in the distance, a cheetah is crouching with her two cubs. She's teaching them how to hunt, all three lying low, waiting for a blissfully ignorant group of Thomson's gazelles, munching the tall grass downwind, to come into range. The sprint, the dodge ... and the gazelle lucks out. The cubs learn what not to do if they want breakfast.

Zebras, wildebeest, lions, hyenas, jackals, warthogs, giraffes, ostrich, eagles, and a host of other animals—they abound on this flat, dry landscape carpeted in late July with knee-high yellow grass. And finally, off in the distance, we see the elephants. A majestic line, matriarchs in the lead, kids in tow, the big males keeping an eye on things. They arrive at the watering hole about the same time we do. In they go, sinking stomach-deep into the cool black mud. One youngster sinks in so far that only his trunk protrudes, moving through the swamp like a slow-motion periscope. On the bank, there's a tussle. Some of the aunts are giving another youngster a hard time, keeping her in line, and mom suddenly has enough of it. Something gets communicated in elephant-speak, and everything settles down.

Those are experiences that feed your soul. They are also experiences that feed the Africans who reap the huge economic benefits of ecotourism—a win-win if there ever was one. In Kenya alone, for example, the loss of elephants—and other threatened species like lions, cheetahs, and hippos—would mean loss of livelihoods for everyone involved in the ecotourism industry. It's not a trivial sum. In dollars and cents, it amounts to about 14 percent, around $10 billion annually, of Kenya's gross domestic product.[3] That is natural capital indeed.

It doesn't take a financial genius to figure out which view of elephants—the bandit view or the ecotourism view—is more profitable over both the short and long term. One of the many problems with the bandit view is that you run out of natural capital pretty darn fast. As we read in chapter 1, if elephants continue to be poached at the rate they were from 2009 to 2013, in as little as twenty years there will be no more wild elephants left on Earth. It's not a business model that works for the long term. The second big problem with the bandit view is the fact that it *is* relatively few people—in this case, outlaws—who are profiting at the expense of everyone else. All said and done, the profits end up as a few million dollars per year for less than twenty years, almost all of which finds its way into a few illegal pockets, and then nothing after that because elephants will have become extinct.

On the other hand, the ecotourism yield from elephants and other threatened species amounts to billions of dollars annually, potentially in perpetuity, distributed through a large sector of the Kenyan economy, with the added benefit of the aesthetic, emotional, and moral gains—not only for Kenyans but for global society—that come from making the economically smart decision, ensuring healthy populations of elephants in their natural setting. Similar examples can be found worldwide. Take these statistics, summarized in a scientific consensus statement presented to world leaders in 2013:

> [In] the Galapagos Islands, ecotourism contributed 68% of the 78% growth in GDP that took place from 1999–2005. Local economies in the United States also rely on revenues generated by ecotourism linked to wildlife resources: for example, in the year 2010 visitors to Yellowstone National Park, which attracts a substantial number of tourists lured by the prospect of seeing wolves and grizzly bears, generated $334 million and created more than 4,800 jobs for the surrounding communities. In 2009, visitors to Yosemite National Park created 4,597 jobs in the area, and generated $408 million in sales revenues, $130 million in labor income, and $226 million in value added.[4]

That kind of projection of dollars generated per year, year after year, and the associated intangible benefits that also accrue, is what natural capital and ecosystem services are really about. And ecotourism is just a small slice of the whole natural-capital pie. In fact, at least 40 percent of the world's economy and 80 percent of the needs of the poor are supplied from biological resources, much of which come from species that are under siege.[5] That siege in almost all cases comes back to valuing short-term, relatively small profits for a few people more than long-term wealth for many. It plays out both species by species and on a broader scale, through the destruction of diverse ecosystems.

In the case of individual species, the short-term profit margin is ultimately rooted in cultural traditions that are a legacy of long-gone times when there was no danger of destroying the species the tradition relied upon. In the elephant example, the cultural tradition is the high value placed on curios carved from ivory, particularly in Asian markets. In the case of many other species, perceived medicinal value drives the trade, a value often related to ills that manifest mainly in the bedroom. Tiger penises, loris tears, pangolin scales, rhino horns—these all fetch high market prices as cures for erectile dysfunction and other sexual problems, despite having no documented effect in curing the ills for which they are prescribed. In the case of erectile dysfunction, consumer choice indicates modern pharmaceuticals are clearly more effective. In Hong Kong, a majority of men over fifty who were users of traditional Chinese medicine switched to drugs such as Viagra to treat their erection problems when given the opportunity.[6]

It's not only treating sexual dysfunction that drives the trade. Body parts of animals threatened with extinction are sold as traditional medicines for a wide variety of ailments. In most cases, the tradition goes back to times well before modern medical principles were known, and there is no proven health benefit. In all cases, there are more effective alternatives. The long list of medicinal uses purported for tigers is illustrative: tiger flesh to treat nausea, malaria, and more than thirty other diseases; ground-up claws to treat insomnia; teeth to bring down fever and treat running sores; fat to treat nausea, dog bites, hemorrhoids, and "scabby, bald-headed conditions" in children; blood to strengthen constitution and willpower; bile to treat convulsions; nose leather to ease irritation from bites and induce birth of boys; bone to treat rheumatism, arthritis, headaches, and dysentery; eyeballs to treat malaria and epilepsy; tail to treat skin problems; whiskers to treat toothaches; brain to treat laziness and pimples; penis as an aphrodisiac; and feces to treat boils, hemorrhoids, and alcoholism.[7] Belief in the curative powers of tigers goes back thousands of years in Asian cultures, and no wonder, given the majesty and power of the animals themselves. Just seeing the fresh paw print of a wild tiger in an Indian forest was awe inspiring to me.

But the reality is, there are only about 3,200 wild tigers left, a mere 3 percent of the number alive a century ago. All belong to a single species, *Panthera tigris,* divided into six living subspecies: the Bengal tiger (*Panthera tigris tigris*), the Indo-Chinese or Corbett's tiger (*P. t. corbetti*), the Malayan tiger (*P. t. jacksoni*), the Sumatran tiger (*P. t. sumatrae*), the Siberian or Amur tiger (*P. t. altaica*), and the South China tiger, also known as the Amoy or Xiamen tiger (*P. t. amoyensis*). That is down from what used to be nine subspecies. The Bali tiger (*P. t. balica*) had been driven to extinction by 1937; the last wild Caspian tigers (*P. t. virgate,* also known as the Hyrcanian tiger or Turan tiger) and Javan tigers (*P. t. sondaica*) died in the 1970s.

Through much of the twentieth century, it was hunting for pelts, or clearing predators from areas developed for agriculture, that decimated tiger populations. Since the mid-1980s, however, the chief threat to the few remaining tigers has been their high value in the traditional medicine trade—simply put, tigers have become worth more dead than alive. Ironically, it was the pre-1980 decimation that helped drive their medicinal price higher—tigers as a commodity became scarce. At the same time, growing human populations followed by increasing affluence in Asian countries significantly increased both the demand and the ability to pay for expensive traditional medicines. Consequently, the profits to be made from poaching tigers and selling their parts on the black market skyrocketed.

That is the situation in which the meager 3,200 wild tigers that now remain on Earth find themselves. They are even closer to the brink of extinction than that number would suggest because effective population sizes are considerably smaller—the world's remaining tigers are broken up into tiny groups sprinkled through several Asian countries and Siberia, with each isolated population of very small size and subject to disappearance if it loses just a few more individuals. For instance, only 600 Amur tigers (*Panthera tigris altaica*) remain in the boreal forests of northeastern China and the Russian Far East, down from 3,000 individuals early in the twentieth century. Population simulations show that if poaching removes only 2 percent (12 tigers out of a population of 600) per year over the next fifty years, there is a 70 percent chance that the population will die out. At poaching rates above 4 percent, the probability of extinction is 100 percent.[8] Those stark numbers make it clear that the days of using tigers for medicinal purposes—dubious, at best, in yielding any health benefits—are numbered. We'll simply run out of tigers.

Rhinoceros find themselves in the same situation. Their horns are a mainstay of Chinese traditional medicine and have been for millennia. The powdered horn is considered an essential ingredient in remedies for diverse maladies: stomach problems, arthritis, rheumatism, melancholia, laryngitis, nosebleeds, rectal bleeding, smallpox, gout, fever, snakebites, and hallucinations, to name but a few.[9] More recently, especially in Vietnam, rhino horn has begun to be used as a "powerful aphrodisiac" and to treat hangovers.[10] Use as a sexual pick-me-up appears to be a recent innovation, but ingesting rhino horn as a sort of Alka-Seltzer to alleviate the aftereffects of a night of binge drinking fits right in with its long-touted properties as an antidote to poisons.[11] Rhino horn has also recently been claimed, through totally unfounded internet rumors, to cure cancer.[12]

When we talk about losing rhinos, we're actually talking about losing five species and four genera, and that is a deep cut into the evolutionary tree indeed. Two species in separate genera used to be common in Africa, the black rhino (*Diceros bicornis*) and the white rhino (*Ceratotherium simum*). Black rhinos aren't really black, and white rhinos aren't really white. The common name of the latter is in fact a transmogrification of "wide-lipped" rhino, referring to the boxy shape of the mouth of *Ceratotherium simum.* "Wide" subsequently became "white," possibly because originally the common name was spoken in Afrikaans or Dutch, in which "wide" is *wyd, wijd, whyde,* or *weit*—easily corrupted to "white" in English. "Lipped" was dropped, leaving the common name white rhino. As for black rhinos, well, they're different from white rhinos, so to clearly distinguish them, they were somehow labeled black. A distinguishing feature between the two is in fact the mouth shape—squarer in white rhinos, more beak-shaped in black rhinos.

It is estimated that at the beginning of the twentieth century, there were more than one million black rhinos in sub-Saharan Africa; in the 1970s that number had dropped to around 65,000, and by 1984, there were only 8,000 or 9,000 left. The species bottomed out at 2,410 in 1995, at which time intensive conservation efforts went into play—which involved, literally, putting the remaining individuals under armed guard. With that level of protection, numbers grew to around 4,880 individuals by 2010—still a 99 percent loss compared to the number of animals a hundred years ago. That drastic reduction led Ronald Nowak, a mammalogist well known for his compilation of information about mammals of the world, to call the black rhino decimation "the greatest single mammalian conservation failure of the twentieth century."[13]

In the past five years, the pressure on black rhinos has once again intensified, because now there is big money to be made. As with tigers, the early culling was largely precipitated by trophy hunting and by attempts to convert land for agriculture—attempts that often failed in the arid lands in which the black rhinos lived. For example, Richard Ellis reported that in the mid-1900s, a single hunter was hired to kill at least hundreds of black rhinos near what is now Tsavo National Park in Kenya, as an attempt to eradicate the tsetse flies that made life difficult for cattle.[14] But, following exactly the same trend as we saw with tigers, over the past few decades the chief driver of extinction for rhinos—black, white, and other species—has become the cold, hard cash that their horns can supply: some $1,400 per ounce as of 2013, about the same price as for gold.[15] At that market value, a single rhino horn can sell for between $50,000 and $150,000, depending on its size. Some reports say a single horn can sell for as much as $300,000.[16]

Rhinos have experienced that kind of pressure since at least the 1970s. By the 1980s, the problems of rhino poaching for the traditional medicine trade were exacerbated by demand in Yemen for an entirely different use of the horns—as dagger handles. Rising incomes provided by Yemeni men working in booming Saudi oil fields meant that many more people could afford to pay top dollar for "jambiyas," the daggers Yemeni boys are given to signal they've reached manhood, and there is no jambiya more prized than one with a handle carved out of rhino horn. That helped lead to a twenty-fold increase in the price of rhino horn during the late 1970s and 1980s. When the oil boom ended at the close of the 1980s, demand in Yemen went down, but overall prices stayed high because the medicinal trade never faltered and rhinos were becoming an even scarcer commodity.

In the past five years, both the medicine trade and prices have increased dramatically. Rising wealth in China and Vietnam has made expensive traditional medicines accessible for many more people, and, in the case of Vietnam, rhino horn has recently become valued for the unfounded, nontraditional uses I pointed out earlier—curing hangovers and cancer, and as an aphrodisiac.

It is not only the black rhino that has been decimated because of this. White rhinos are faring a bit better but are still on a steep downhill trajectory. In fact, the northern subspecies (*Ceratotherium simum cottoni*) is all but gone. In 1960, about 2,400 individuals were thought to exist, mostly in Garumba National Park in the Democratic Republic of the Congo (DRC), but the numbers dwindled after that, and between 2003 and 2005, poachers killed about half of all the northern white rhinos in

Africa. By 2006, surveys found only four northern white rhinos left in the wild. Now there are none except for four that have been moved from a Czechoslovakian zoo to a closely guarded nature reserve in Kenya, where the hope is that they will produce offspring.[17] The other white rhino subspecies is called the southern white rhino (*Ceratotherium simum simum*) to reflect its distribution primarily south of the DRC, in Kenya, Uganda, Zambia, Mozambique (until recently, anyway), Zimbabwe, Botswana, Namibia, and South Africa. The subspecies was nearly eliminated by slaughter for its horns in the late nineteenth and early twentieth centuries. In the only rhino conservation success story to date, reintroduction of the southern white rhinos into strictly protected nature reserves all over southern Africa brought their numbers back up to a little over 20,000 by the year 2000.

However, since 2007, increased poaching has begun to bring their numbers down again. In South Africa, a stronghold for the subspecies and for rhinos in general—it holds about 83 percent of the African individuals and 73 percent of all rhinos worldwide—about fifteen rhinos per year were poached from 1990 to 2007. Beginning in 2008, the number of rhinos found shot and with their horns hacked off increased every year—from 83, to 122, and so on, until the death toll in 2012 was 688. Even worse was 2013—in the first quarter of the year, 188 South African rhinos were killed by poachers, with 135 of them shot in Kruger National Park, one of Africa's crown jewels.[18] By May 16, the death toll had risen to 313; by the end of the year, 1,004 poached rhinos lay dead.[19] That rate of loss means that by the year 2016—just three years from when I'm writing this—the number of deaths will start to outweigh the number of births, and soon after that, the southern white rhino will be a thing of the past.[20] Outside of South Africa, the losses are every bit as dramatic. The last of the 300 southern white rhinos introduced into Mozambique for protection in 2003 were killed in April of 2013.[21] In Kenya, southern white rhino killings in 2013 were projected to be double those of 2012, with 7 shot dead in just one week in May, all in what were supposed to be rhino sanctuaries.[22]

In Asia, rhinos are all but gone, even though that was once the rhino hot spot as far as diversity of genera and species goes. The Indian rhinoceros (*Rhinoceros unicornis*) is doing the best but is represented by fewer than 3,000 individuals, more than 70 percent of which are in a single national park (Kaziranga National Park in the state of Assam in northeastern India).[23] Like the southern white rhino in Africa, Indian rhinos in Kaziranga initially were a conservation success story, their population growing from 12 animals in the early 1900s to some 2,300 due to effective protection by Indian park rangers and managers. But as with the southern white rhino, poaching is now starting to erase a century's worth of conservation successes. In 2011, 11 Indian rhinos were killed by poachers in Kaziranga; in 2012 the number rose to 38; and in just the first four months of 2013, at least 16 were slaughtered and dehorned in and around the park.[24]

That is a rosy picture compared to that of the Sumatran rhino (*Dicerorhinus sumatrensis*). In 2008, only 250 were left, down by 80 percent over the previous three rhino generations (totaling about sixty years), and the numbers were expected to drop by 25 percent over the ensuing twenty years at an average of a little over 3 rhinos per year, given then-current poaching rates.[25] That estimate turned out to be wildly overoptimistic. As of April 2013, only five years later, poachers had wiped out well over half of the animals that were alive in 2008, leaving only 100 Sumatran rhinos in the wild.[26]

In even more dire straits is the Javan rhino (*Rhinoceros sondaicus*). Formerly spread through a wide swath of Bangladesh, Myanmar, Thailand, Laos, Cambodia, Vietnam, peninsular Malaya, Sumatra, Java, and probably southern China, the only place they are left—just 60 individuals—is the western tip of Java.[27] Up until 2011, a few individuals lived in Vietnam, but poachers there got them all.

Tigers and rhinos are but two of the species being driven to extinction because of their value for traditional Chinese medicine (as well as for nontraditional remedies, in the case of the rhino). The list of other such species, both animals and plants, is long, as one can sense from this excerpt from an article in the conservation journal *Oryx,* published in 2004:

> Amongst the 42 animal species that in 1995 were banned from domestic markets by the State Traditional Chinese Medicine Administration Bureau and the Ministry of Public Health in China are . . . three species of rhinoceros . . . tiger *Panthera tigris,* leopard *P. pardus* and snow leopard *Uncia uncial,* the musk, horns, tail and penis of three species of musk deer (*Moschus* spp.), the scales of all seven species of pangolins (*Manis* spp.), the antlers, velvet and horns of a number of species of deer and antelope, the carapace of three species of turtle, the testes and penis of all species of seals, and dried seahorse *Hippocampus kelloggi.*[28]

How to eliminate the killing of species that is spurred by the black-market trade for purported medicines (and, in the case of elephants, for curios) is one of the thorniest issues we face in dodging the Sixth Mass Extinction, because deep-seated cultural traditions make people willing to pay exorbitant prices for "cures" that usually do nothing to enhance health or, at best, have no more beneficial effects than products provided by non-threatened species or modern pharmaceuticals. One important part of the solution lies in educational efforts aimed at forcefully conveying that the very promotion of those particular cultural conventions will soon kill the traditions, because the species they rely on will be gone.

That message is beginning to be heard by some consumers of threatened species, and it seems to be most effective when communicated widely through social media by celebrities well known to those in the cultures that currently are driving the black market. For example, Asiatic black bears (*Ursus thibetanus*) are highly prized in traditional Chinese medicine, especially for their bile and gall bladders. Their numbers have declined by nearly 50 percent over the past thirty years and are expected to continue on that downward spiral if current pressures—habitat destruction and their use in traditional medicine—continue.[29] The bears are either killed outright and sold for parts or trapped and held in captivity so that their bile can be "milked" (a procedure that generally involves permanently implanting a catheter in the bear's gall bladder and confining the bear in a small cage).[30]

The twist to this story is that Guizhentang Pharmaceutical, China's largest producer of bear bile, was poised to ramp up its operations by submitting an application to go public on the Shenzhen Stock Exchange. What Guizhentang didn't count on was basketball star Yao Ming and pop star Han Hong mounting a social media campaign outlining, in no uncertain terms, why that was such a bad idea. As a result, Guizhentang Pharmaceutical withdrew its application. Almost overnight, the economic venture became a nonstarter.[31] The power of promoting cultural change in this way is immense—there are now more than 3.2 billion people worldwide linked by 6 billion mobile phone connections, most of them utilizing text messaging. Nearly one-third of the world population uses the internet, and in China, that number is even higher, nearly 40 percent. Similar campaigns to educate the consumers of products that require killing elephants, tigers, rhinos, and other species will be an essential step in keeping those species alive.

Social media campaigns tend to be most effective with people born into the age of the internet, that is to say, the generation whose members are just now beginning to assume leadership positions. Awareness of, and action to mitigate, environmental issues is growing among that demographic, as illustrated by the response to the bear bile campaign. All this is good news in promoting hope for the future, but the other part of the reality is that the demographic currently fueling the market for traditional medicines, generally the over-thirty crowd (mostly way over thirty), is not likely to either change its views or go away in time to save the few remaining individuals of the species most at peril.

That's where fighting fire with fire comes in: that is, treating the poaching of threatened species as a war, which is exactly what it is, especially given that in many cases the profits go to supporting international terrorist operations and crime rings.[32] Whether the good guys or bad guys will win this one is still up for grabs. In order to win, the good guys will have to get a lot more support from international and local communities in the form of money, weapons and other logistical support, strict laws, and follow-through on stiff penalties for offenders.

So far, the good guys have employed primarily defensive tactics, such as the obvious approach of stepping up patrols of heavily armed rangers in areas where the at-risk animals live. The economic cost of this is small, especially by developed-country standards—less than $5,000 per ranger per year.[33] Employing twenty-five or so rangers to protect a rhinoceros or elephant costs less than what a single rhino horn or elephant tusk from that animal could be sold for at today's prices. Therefore, increasing funding—through either private or governmental sources—to provide armed protection of at-risk wildlife is perhaps one of the most cost-effective means of reducing poaching losses.

Of course, more is at stake for those rangers than money, much more: they sometimes lose their lives. We need to up the ante equivalently for poachers and those they work for. As it stands now, the penalties for poaching and smuggling are light, making it well worth a poacher's while to take great risks, and the rangers are usually outmanned and outgunned. An offensive approach, in addition to beefing up the troops on the ground that are trying to prevent the slaughter, would be to aggressively aim for the heart of the problem: the terrorist groups and international crime rings that reap almost all the benefits.

As in any war these days, advanced technology also plays an important role. Among the high-tech approaches currently utilized are unmanned aerial drones that follow movements of wildlife and alert authorities when poachers are getting near; various tracking devices on wildlife (collars, embedded GPS chips, and so on) that constantly send signals to Google Earth maps, which in turn alert rangers to unusual movements that would indicate trouble; fences surrounding wildlife preserves integrated with alarm systems that notify officials when the fences are crossed; and hidden cameras.[34]

There are also experimental programs to make the wildlife products unsalable, most of which are being employed for rhino horns. One program injects the horns of living rhinos with a dye that is easily detected both visually and by airport scanners, and sometimes also with a substance that will cause discomfort to anyone who ingests the powdered horn.[35] The injection procedure involves shooting the rhino with a dart to knock it out for the short period of time (usually less than an hour) it takes to drill a couple of small holes in the horn, through which the dye is then pumped; the rhinos seem not to be harmed in the process. Another approach is to dehorn the rhino entirely, rendering the animal much less valuable to poachers. However, poachers have been known to kill dehorned rhinos for the horn stub, or simply out of spite.

Finally, there are some who advocate flooding the market for animal parts by actually "farming" some of the affected species in order to supply the traditional medicine trade. The argument is that legalizing and controlling the market would make it unprofitable to kill wild animals. Rhino farming is perhaps most practical in this regard, because rhinos regrow their horns after they are cut if the horn removal is done properly. In fact, rhino farms already exist in South Africa, and the horns are being stockpiled in hopes that the trade is legalized at some point in the future, at which time the stockpiles could be used to saturate the market and bring down prices.[36]

The downsides are that there is presently no easy way, short of DNA testing at every step of the supply chain, to track whether products that end up being sold in the marketplace come from farmed animals or wild ones; that it would still cost more to set up a legitimate wildlife ranch than to simply go out and kill a wild animal, so poachers would continue to make easy money (though not as much); and that it will be extremely difficult to stop the poaching without changing the psychology of the consumers who are demanding the products.[37] Even more serious, in the case of animals like tigers (which must be killed to render their products economically useful) and bears, it is simply not possible to raise enough animals in captivity to fuel the trade, in part because they don't reproduce well except under natural conditions. Their generation and gestation times are too long, and their present numbers too few, to make a legitimate business of it. That means that the only way to keep captive stocks viable would be to continuously capture animals from the wild and pen them up—pretty much the same as killing them, as far as viability of a species goes. For those reasons, tiger and bear farms, which already exist legally in China, are nonstarters for species conservation.[38] This is not to mention the ethical and moral considerations of keeping such animals in captivity, which in fact was one of the chief drivers of keeping bear bile off the Shenzhen Stock Exchange.

The species-by-species problems that arise from the illicit traditional medicine trade illustrate the even larger, more diffuse extinction problem that comes from going after short-term profit at the expense of long-term wealth. The mounting cost, in ecosystem services, of the current extinction crisis goes far beyond losses of charismatic wildlife. For example, commercial fisheries in the United States, some of which rely on species in which the majority of populations have already gone extinct, provide approximately one million jobs and contribute $70 billion annually to the United States economy.[39] And that is just a drop in the bucket of the economic benefits we get from keeping a diversity of species and their ecosystems alive.

The flip side of the medicine story illustrates this nicely. Lest you get the idea from what you've been reading up to now that the medicinal value of species is for the most part quackery, think about where perhaps you and almost certainly people you know would be without the jungles of Brazil, Paraguay, and Argentina and a very deadly snake that lives there, the fer-de-lance viper. You or those loved ones might very well be dead, if you (or they) are like many people in my family, including my mother. In her late fifties she was diagnosed, like millions of other people, with high blood pressure, which goes along with related heart problems that happen to be the chief cause of death for the human species. More than seven million of us die annually from those cardiovascular ailments, and related cerebrovascular problems take another six million. My mother lived to the respectable age of ninety—an event suitably celebrated at a party where the bourbon flowed freely and her eighty-something brothers serenaded her with Slovenian songs—but she very likely would have died of a heart attack long before that without the right medication for her hypertension, a little pill she took each day called an ACE inhibitor.

ACE is short for angiotensin converting enzyme, which is important in a string of chemical reactions that starts with the kidney producing a substance called renin, which in the bloodstream is converted to angiotensin I. With the addition of the ACE, angiotensin I is converted to angiotensin II, which constricts the blood vessels and consequently raises blood pressure, not a good thing for people afflicted with hypertension. ACE inhibitors stop that conversion, keeping blood vessels relaxed and blood pressure low.

And that's where the fer-de-lance viper, the most unexpected kind of natural capital, enters the story. Its deadly venom provided the breakthrough for developing ACE inhibitors. The venom is so deadly, in fact, that just a few drops of it on the arrowheads used by indigenous Brazilian tribesmen would bring down a large animal as long as the hunter's aim was good enough to draw blood. The venom contains a compound that breaks down blood vessel walls, causing the wounded animal to systemically bleed to death. Knowing that, and suspecting the hemorrhagic compound in the venom might disrupt the ACE progression in people, in the 1960s a young Brazilian postdoctoral student named Sergio Ferreira brought some of the venom to John Vane's lab at the Royal College of Surgeons in England. There, they determined experimentally that the compounds in fer-de-lance poison would inhibit angiotensin I from converting to angiotensin II—they had found the first ACE inhibitor.[40] Subsequent biochemical refinement of the molecule led to a compound that could be taken orally, which became the key ingredient in Captopril and later derivatives. Since being rolled out in 1975, Captopril and related drugs have prolonged millions of lives and made the Squibb pharmaceutical company (now Bristol-Myers Squibb) hundreds of millions of dollars.

The lives saved and money generated didn't stop there. Later, Captopril was shown to be effective in reducing the rate of a certain type of kidney failure. Economic models showed that with the use of Captopril to treat kidney failure, the savings in medical expenses over a ten-year period would amount to $2.4 billion.[41] Tens of billions more dollars are added to the world economy when you take into account the extended productivity of workers whose health is improved by Captopril.

The point here is that natural capital is all around us, but we usually don't acknowledge what it gives us. In the Captopril example, the natural capital is what most people view as a deadly snake, best to stay away from or get rid of. Yet without it, millions of people would have a shortened lifespan and billions of dollars now rolling through the economy would have been lost. Captopril is far from an isolated example. Most major pharmaceutical companies are involved in developing new drugs from a variety of species, from plants found in rainforests to snails found in shallow marine waters, and small wonder: life-saving medicines derived from plants and animals yield about $43 billion annually.[42] And they are but one kind of natural capital that provides essential ecosystem services—both for sustaining life and making it more pleasant—which would be lost without intact ecosystems that harbor a reasonably full complement of native species.

Other examples of ecosystem services include stabilizing water supplies and filtering drinking water; protecting agricultural soils and replenishing their nutrients; pollinating crops and wild plants; providing food from wild species (especially seafood); disposing of wastes; controlling the spread of pathogens; moderating weather; and helping to reduce greenhouse gases in the atmosphere.[43] And, for the most part, more diverse ecosystems—that is, those in which species have not gone extinct or otherwise been deleted—tend to provide a greater variety and higher quality of ecosystem services.[44]

When you begin to recognize the diversity of essential services that ecosystems provide, they can be economically valued in myriad ways. For example, in the case of a rainforest, it is often tempting to cash in immediately by cutting down all the trees, making a quick buck by selling the timber, and then converting the clear-cut area into a soybean field, palm plantation, or cattle ranch. If ecosystem services are valued, though, that is exactly the wrong choice. Once that rainforest is cut down, some of the ecosystem services that are lost forever include storing CO_2 (which slows costly climate change considerably); maintaining fertile soils (in the absence of the rainforests, which effectively cycle nutrients, soils rapidly become infertile); and regulating climate (the respiration of rainforest plants produces up to 50 percent of the rainforest's rain and maintains the local water cycle).

The short-term versus longer-term tradeoffs may seem unimportant at first, but they become crystal clear when you take the time to do some math. The economic costs of the short-term gain get very big very fast, especially when you project them out a few years over wide regions. For example, deforestation and other land transformation worldwide account for around 20 percent of the CO_2 humans add to the atmosphere annually, which roughly equates to about 0.4 degrees Fahrenheit (0.2 degrees Celsius) of the overall warming projected by the year 2050 if we remain on the path we're on now.[45] As I write this, the dollar value of keeping that CO_2 out of the atmosphere ranges from about $6 to $60 per metric ton.[46] The World Land Trust calculates that a typical acre of rainforest sequesters 75 to 109 metric tons of CO_2 over twenty years, or about 4.6 tons annually. Thus the value of leaving the trees standing amounts to about $138 per acre per year.[47] Given that about 38 million acres of rainforest are cut down each year, the loss in carbon sequestration amounts to about $5.24 billion per year, or about $105 billion over twenty years.[48]

Contrast that with the money to be made clearing 38 million acres of rainforest each year for twenty years and planting it in soybeans. We can get a ballpark figure using prices and costs that are currently representative for Brazil. One acre of cleared forest can yield, on average, fifty bushels of soybeans. Brazilian soybean prices fluctuate widely through the year—for example, from $17.34 per bushel in September 2012 to $11.11 per bushel in May 2013—but let's use a reasonable number, say, $15 per bushel on average.[49] Subtract about $10 per bushel to cover the cost of growing and transporting the soybeans to market and the yield per acre is about $250.[50] Compare that with what you could make by not cutting the trees and charging for their value in carbon sequestration instead—about $138 per acre—and suddenly soybean farming looks much less attractive. You can make more than half of what you would earn by working your butt off by simply sitting back in your hammock and selling the carbon credits. Now, what if the average price of carbon rose to $60 per ton, which is the value ExxonMobile placed on it in 2013? At that point, simply watching the forest grow earns you $26 *more* per acre than planting soybeans. And remember, this accounts only for the carbon credits—it doesn't take into account all those ecosystem services that we have not yet put a dollar value on, like the potential losses in pharmaceuticals, water filtration, and recreational opportunities. The total value of all ecosystem services is very hard to get a handle on, but an early estimate, published in 1997, was about $33 trillion per year for the global total, which at that time exceeded the entire world GDP (roughly $30.5 trillion).[51]

If ecosystem services are so valuable, why aren't they firmly integrated into our economic system? One part of the answer is what we saw with the poaching examples: people tend to think in terms of short-term gain rather than long-term wealth, and they tend to hang on to old ways of doing things. The other part of the answer is that biologists and economists have only recently begun figuring out ways to put a dollar value on ecosystem services and how to build viable business models around those services. Valuing and trading in ecosystem services have been advancing rapidly, and that is good news for keeping species and diverse ecosystems alive and well. In fact, individuals, businesses, and governments are beginning to use business models that integrate the value of ecosystem services in their day-to-day decisions.

Trading in carbon credits is but one example of how such business models are beginning to help people from the local to the global level.

Say you own or otherwise have rights to a few acres of land in a Brazilian forest. If you needed to make some money off of that parcel, until recently you wouldn't have had the option of getting paid to leave the forest standing. Even now, that option does not exist everywhere, but the door is opening through programs such as REDD, Reducing Emissions from Deforestation and Forest Degradation. Basically, REDD provides a mechanism by which international agencies such as the United Nations, nongovernmental organizations such as the Environmental Defense Fund, and national and local governments cooperate to make it possible for the guy on the ground to choose between getting paid to cut his rainforest or getting paid to leave it standing.[52] As the quick calculations I went through above show, with fair carbon pricing, it would be more profitable to leave the forest standing, but this has seldom happened because few people in the backwoods of Brazil or the other countries that harbor intact, highly biodiverse forests have the foggiest idea how to get into carbon trading. That's where programs like REDD come in—the goal is to establish the infrastructure, educational programs, and regulatory procedures needed to enable local people to profit by leaving forests standing.

Carbon trading so far is a key focus of the REDD program, and because that is such a controversial topic, REDD engenders its share of controversy as well. The basic idea behind carbon trading is that industries and individuals who emit more than their fair share can offset their emissions by paying someone else to emit less. Setting that "fair share" is one of the key points of contention, in no small part because it implies imposing a tax on heavy emitters. Another big stumbling block is exactly who should be allowed to reap the benefits. So far the REDD program is restricted to developing countries—the lowest carbon emitters—but does that mean that a multinational company can buy up rainforest in a developing country to offset its polluting activities elsewhere? Or should regulations be designed to favor local citizens? These are among the difficult issues that make programs like REDD works in progress, but the fact that they are even on the table is encouraging.

Also encouraging is that big corporations are getting into the environmental awareness business. In part, that's because consumers are sending clear signals that given the knowledge and the choice, they prefer businesses that are environmentally aware. It is that awareness, not a carbon tax, for example, that drives Siemens Healthcare to partner with the World Wildlife Fund and to advertise that it is reducing CO_2 emissions by extending the useful lifetime of its products and replanting trees in logged Indonesian forests with every refurbished piece of medical equipment it ships.[53] Likewise, the giant agro-company Cargill, following campaigns by the eco-activist group Greenpeace, emplaced a moratorium on buying soybeans that came from Brazilian land that was deforested after 2006. Cargill also pressured other major soy buyers to do the same, and the moratorium remains in place as of 2013, a story that is related in detail in Mark Tercek and Jonathan Adams's recent book about how environmental awareness is beginning to work its way into the DNA of big business.[54]

Tercek (a former investment banker and in 2013 the CEO of the Nature Conservancy) and Adams point out an even more compelling reason that big business is getting greener: there is a growing recognition that natural capital and accounting for ecosystem services are good for the bottom line. It's for that reason that the Coca-Cola company, worth about $167 billion and generating a cool $8.5 billion in profits for its shareholders in 2012, now specifies water as natural capital that affects its profits right up front.[55] After one of its main plants was shut down because of a lack of water in India, with a high court ruling that residents could seek compensation from the company for depletion of groundwater between 1999 and 2004, Coke emplaced a program to become "water neutral"—that is, to ensure that the amount of water taken from nature and communities is balanced by the amount of water returned. Moreover, Coca-Cola listed water as natural capital (raw material) in its 2010 Securities and Exchange Commission filing: "Our company recognizes water availability, quality, and sustainability ... as one of the key challenges facing our business."[56]

Likewise, governments at all levels are increasingly recognizing how ecosystem services figure into their economic growth and long-term viability. What's going on in China right now demonstrates that. China learned about the costs of deforestation the hard way, which is to say, the expensive way. With wholesale clearing of forests came damaging floods in the Yangtze region, which in 1998 killed thousands of people, destroyed millions of households, stripped the topsoil, and clogged rivers with sediment. That drove home the need for the most massive reforestation project in history, and now 120 million farmers are being paid to farm using methods that stabilize steep slopes, control floods, and maintain biodiversity while at the same time ensuring successful yields of crops and timber.[57] In a similar vein, Costa Rica used to have some of the highest deforestation rates in the world, but emplacement of a national payment system for ecosystem services changed that to one of the lowest rates.[58] The trick now is to install those payment-for-ecosystem-services plans globally, such that it no longer makes economic sense for a country that uses a lot of ecosystem services (like China or the United States) to protect the needed resources within its own borders, only to destroy them in other countries. "Not in my backyard" attitudes are responsible for much of global deforestation and loss of natural capital (for instance, the international market for timber and palm oil, rather than local pressure, is driving the decimation of Amazonian and Indonesian rainforests).

Emplacing financial plans that view nature as a long-term investment takes out-of-the-box thinking that doesn't neglect the bottom line. Not surprisingly, accounting for and protecting ecosystem services usually turns out to be the most economical way to go. In New York City, the municipal water authority has found that investing in natural landscapes for water filtration is more economical than building filtration plants.[59] That sort of financial acuity has been scaled up even more in Latin America, where cooperation between nongovernmental organizations and municipal governments has built up "water funds," investment funds that generate interest that's used for projects that contribute to the long-term viability of entire watersheds, thereby maintaining important ecosystem services.

The idea behind water funds is compelling and effective: they provide for the people who need clean water to drink, for those who need water to produce goods and services, for others who need to make a living off of the natural resources, and for biodiversity conservation. Basically, municipalities, companies, and people who rely on the water pay into a fund in exchange for the ecosystem service they receive—clean water. The fund generates income, which is then used for things that maintain or improve water quality. The interesting thing is that the return on investment can be enormous. An example is Ecuador's Quito Water Fund. The money came from modest investments by philanthropic organizations and the government; as water users made voluntary contributions, the fund grew until, as of 2012, it was generating about $800,000 annually. That money is used to keep the watershed healthy. As a result, the Quito water company, just like New York City's, has saved lots of money because it didn't have to build costly filtration plants; instead, it has been able to rely on natural filtration that takes place in healthy watersheds. Why have those watersheds remained healthy? Because the spin-off from the water fund provides money to upstream residents for things like community-based reforestation, building fences to keep livestock away from streams, and funding micro-enterprises to provide livelihoods for people who otherwise would need to farm in ecologically destructive ways.[60]

As a result, biodiversity also benefits. In the case of the Quito area, important Andean condor habitat is being preserved. With a population of only about sixty-five birds in Ecuador in 2002, and those separated into disjunct populations, Andean condors (*Vultur gryphus*) are a declining species but are among the most impressive birds in the world for, among other reasons, their sheer size.[61] Their wingspans can measure over ten feet from tip to tip. Exactly how big that is was put into perspective one austral spring when our then four-year-old daughter picked up an Andean condor wing feather she found lying on the ground; the feather was about as tall as she was.

The win-win model illustrated by the Quito Water Fund has another important attribute: it is scalable. As of 2012, at least thirty-two water funds were in various stages of development across Latin America, through the collaborations of what used to be unlikely bedfellows: environmental organizations like the Nature Conservancy, corporations like Dow (the chemical company) and Rio Tinto (the mining company), and governments like that of Colombia. The general concept is proving sound: investing in nature to decrease long-term capital costs for companies and municipalities can have the added attractions of enhancing both biodiversity and livelihoods.

That last point is a key one—people need to make a living, so demonstrating the value of ecosystem services and natural capital at the level of individual enterprises is essential. In some cases, that value is obvious—ask any almond grower in California about the cost of using bees to pollinate his or her crops. Over the past few years, colony collapse disorder, probably caused by some combination of a spreading virus and pesticides, has been wiping out bees in record numbers.[62] Farmers can't count on the bees hanging around naturally to do the work of pollinating their crops; there are just too many crops. Instead, they rent hives from beekeepers who travel the country, following the flowering of various crops in various places. For California almond growers, that happens to be January and February, when, in 2013, the going rate for bees climbed to more than $200 per hive, almost four times what a hive cost in 2003–2004.[63] That cost inevitably gets passed on to consumers.

In other cases, the added value of ecosystem services is subtler and takes some fairly detailed analyses to quantify. Such studies are not very numerous so far, but those that have been done indicate that when you lose natural capital, you lose money. For example, a team of conservation biologists composed of Taylor Ricketts of the World Wildlife Fund, Gretchen Daily and Paul Ehrlich of Stanford University, and Charles Michener of the University of Kansas Natural History Museum demonstrated that Costa Rican coffee farmers whose fields are near natural forests are apt to make an extra $60,000 per year.[64] The ecosystem service providing that income boost comes from the natural pollinators in the forest (in this case, feral and native bees, not bees from beekeepers' hives), the proximity of which ended up increasing both coffee yields and quality.

If natural capital and the resultant ecosystem services are so valuable to producing revenue, why have we not already integrated them into our economic thinking? In part it's because people have tended to view natural capital as an unlimited resource. Now we know that is no longer the case, as we run up against limitations in such basic necessities as water, forests, and fish, not to mention the many other thousands of species whose decimated populations put them on the edge of extinction. But another very important reason is that we just haven't known how to cost out the value of nature.

That problem is now being solved as well. A new breed of economists and biologists is beginning to calculate the value, in dollars and cents, of many of these essential ecosystem services, and to develop ways to feasibly integrate them into the global economy. Importantly, these new research initiatives are producing internet-based tools and how-to guides that allow corporations, governments, and individuals to plug in a few critical pieces of information to answer questions like "How will a new coastal management plan impact seafood harvest, renewable energy production, and protection from storms?" or "Where would reforestation or protection achieve the greatest downstream water quality benefits?" or "Which parts of a watershed provide the greatest carbon sequestration, biodiversity, and tourism values?" One tool that does that, called the Integrated Valuation of Environmental Services and Tradeoffs, or InVEST, is a software package that maps the economic value of ecosystem services using alternative scenarios for both terrestrial and marine areas.[65] It came online as part of the Natural Capital Project, a collaboration of cutting-edge biologists and economists at several universities and nongovernmental organizations. It has already been used with notable success, for example in helping the largest landholder in Hawaii, Kamehameha Schools, design a land-use plan that balances values from multiple stakeholders against environmental health. In that application, InVEST demonstrated that the best choice was a plan that emphasized diversified agriculture and forestry, which resulted in a positive financial return of $10.9 million.[66] Several other ecosystem valuation tools and data resources, some for specific ecosystems, have recently become available.[67]

All of this could bode well for saving species in the coming decades. As the price of ecosystem services is factored into the costs of more and more financial transactions, it is becoming apparent that where natural capital is involved, a better bottom line for the long term results from keeping more species alive and diverse ecosystems healthy. That important realization is beginning to reach more and more businesses, and in addition, more and more consumers are beginning to favor products that are environmentally friendly. The open questions, of course, are whether these nascent realizations will lead to a paradigm shift in how people think about making and spending money, and whether that will happen before it is too late for many species. For the people dead set on killing elephants, tigers, and rhinos or fishing every last tuna out of the sea or cutting down nonrenewable forests, the paradigm shift won't happen out of good will. In those cases, it will take public outcry, strong government regulations and incentives, and severe, strictly enforced penalties for offenders. For businesses that rely on ecosystem services unknowingly, both mom-and-pop operations and multinational corporations, it will take demonstration that the business dies if the natural capital dies, and it will require incentives that put cash in the pocket for practices that maintain, rather than degrade, ecosystem services. For most of us, though, the action is a lot more straightforward. It will simply take deciding that we value species alive more than dead and communicating that decision through how and where we choose to spend our money.

CHAPTER SEVEN

Resuscitation

Let's say, for the sake of argument, that all of the solutions suggested in the previous few chapters have been implemented, and we stop there. Extinction rates would slow, but we still wouldn't be out of the woods. That's because the extinction train has already gained tremendous momentum over the past couple of centuries. Many species are already handicapped by vastly reduced numbers, substantial loss of genetic variation, and loss of most of their natural habitats. And as much as we might like to, we can't forget about the components of that "perfect storm" of extinction that we've already put in place—rapidly progressing changes in the atmosphere, climate, and oceans, and intense new ecological stressors. All of these add up to an extinction debt, which means that even if the contributing causes are stabilized, many species could fizzle out simply because they can't, on their own, recover from the losses already inflicted on them, or because they live in a world that's changing more rapidly than they are equipped to handle.[1] That means people will have to step in to actively help—in some cases actually resuscitate—species that won't make it on their own, and that's going to take some bold new science, of which there is no shortage these days. In fact, the relevant fields of conservation biology and molecular biology have been advancing so fast in the past couple of years that the real problem may be figuring out which of the multiple paths currently being pursued is the one to put your money on.

Conservation biology is the branch of science devoted to protecting species and ecosystems from extinction. Typically, conservation biologists are in the field with the species they are trying to save, and they work with local people and governments, as well as with international organizations, to implement policies that will relieve extinction pressures on species in peril. Molecular biology is concerned with figuring out things like how to modify genes, or how biochemical reactions at the molecular level translate into something that keeps an organism alive. That world is the world of laboratory benches, beakers, pipettes, and replicate experiments—and, these days, it's the world of "de-extinction," which some view as a revolutionary and helpful techno-fix, and others, including most conservation biologists, are up in arms about.

The concept of de-extinction has been around for a while, though the word itself just came on the scene in 2013. I first got an inkling of its popular appeal a year earlier, when I was teaching a freshman seminar about the Sixth Mass Extinction. These seminars are one of the joys of university teaching. The students are excited to be away from home for the first time, on their own at last. They're getting to know a new group of friends and figuring out what college life is all about, and they are happy to be in a class they are taking out of pure interest, not just to fulfill a requirement. So in the best seminars chatter runs high and uninhibited ideas fly around the room.

One morning, the uninhibited idea stopped me dead in my tracks. "Why are we so worried about extinction?" one of the kids asked. "Why not just deep-freeze a few individuals of the species we know are in danger, then, when we're up to the task, we could use their DNA to clone them and bring the species back to life?"

That kid was no intellectual slouch—he'd been reading up on things I hadn't assigned. At the time, there had been talk for years of cloning mammoths, using DNA from their frozen tissues, which are fairly commonly found in permafrost in Siberia. Headlines such as "Japanese

Scientist to Clone Woolly Mammoth within Five Years!" were easy to find on the internet, and to read them, you would think it was just a matter of time before it happened.[2] The idea was to find frozen mammoth tissue with viable DNA and then insert that DNA into an Asian elephant egg cell from which the elephant DNA had been removed.[3] Then the genetically modified elephant egg cell would be implanted in an Asian elephant's womb, and presto, about 645 days later, the elephant gives birth to a baby mammoth. Extinction problem solved.

As you might suspect, it's not that easy, especially for species that have been dead thousands of years, like mammoths. In part, this is because even though their tissues have been in the deep freeze for more than 10,000 years, the DNA is inevitably very degraded. It turns out that DNA is a particularly fragile molecule. Even under the best circumstances it is only partially preserved, and the older the specimen, the worse the preservation. Even in the best cases, especially for tissues thousands of years old, the sequences that survive are typically only 100 to 300 base pairs long.[4] A base pair is a single pair of matched amino acids; thousands of base pairs are strung together to form genes, and the entire genome of animals like mammals and birds contains billions of base pairs. For a visual image, imagine two very, very long strings of beads, twisted together, where each bead touches its neighboring bead on the opposite string: each pair of touching beads is a base pair. Now take some scissors, randomly cut the twisted strands and throw away lots of the beads, then scatter the remaining fragments. What you are left with is analogous to the few strands of DNA, each of just a few hundred base pairs, that are found in the tissue of dead animals—even in recently dead animals, let alone those that have been frozen for over 10,000 years. Because of that, the best that could be hoped for in a "cloned mammoth" would be something that is part mammoth; the other part would be genetically engineered by splicing in pieces of elephant DNA to fill in the missing bits of the mammoth genes. Who knows what such an animal would turn out to be like, assuming there was even live offspring at the end of the process, but it certainly wouldn't be a replica of the species that roamed the tundra during the ice ages. That's why, back in the winter of 2012, when I was teaching that class, reconstituting extinct species from preserved DNA still seemed like the stuff of science fiction, and that was pretty much the answer I gave my student, with the obligatory comparison to the movie *Jurassic Park.*

I felt I had been a little curmudgeonly a year later, when the idea of resurrecting extinct species suddenly took off, both in scientific circles and in the popular press. Almost overnight, re-creating extinct animals was dubbed "de-extinction" and entrepreneurs, molecular biologists, bioethicists, and conservation biologists were debating its pros and cons. The idea hit the mainstream through the efforts of Stewart Brand and Ryan Phelan. Brand is a visionary who, among many other accomplishments, founded the *Whole Earth Catalog* and booked the Grateful Dead for one of their first San Francisco gigs in the 1960s. He is a master of making things happen. Phelan made her name as an entrepreneur in the arenas of medical information, DNA, and biodiversity. She was the founder and CEO of DNA Direct, which provided personalized genetic testing and decision support for patients and their health care providers, and in 2002 she cofounded the ALL Species Foundation, which had the goal of discovering all species on Earth.

Married, Brand and Phelan are a formidable team, as their promotion of de-extinction illustrates. By 2012, they had rounded up a group of scientists with both demonstrated enthusiasm to re-create extinct species and demonstrated expertise in molecular biology and cofounded, with Phelan as executive director, the Revive and Restore project under the auspices of the Long Now Foundation (which Brand also founded).[5] They then worked with National Geographic and the widely acclaimed TED organization to orchestrate a conference on de-extinction, timed to coincide with the publication of an issue of *National Geographic* that promoted the topic and explored its feasibility.[6]

Listening to Brand's TED talk about why we should pursue de-extinction, it's hard not to be convinced. It sounds so, well, cool and doable. The cool factor can't be denied—it would be tremendously exciting to push the frontiers of molecular biology to the extent it would take to make de-extinction a reality, and who wouldn't want to see an extinct species brought back to life? Also, there's a certain amount of truth in how Brand sells the concept: "Last century, discovery was basically finding things, and in this century, discovery is basically making things."[7] In that context, engineering our way out of the Sixth Mass Extinction begins to look pretty attractive. But could it really work?

The answer to that question—which is in essence the same one my freshman student asked earlier—emerges from studying what is now the Revive and Restore project's number one candidate for de-extinction, the passenger pigeon (*Ectopistes migratorius*). In the grand scheme of things, passenger pigeons (known in some literature as wild pigeons) have not been extinct all that long. The last one known even had a name, Martha; she died in the Cincinnati Zoo on September 1, 1914. Martha was the lone remnant of a species that had been incredibly populous in at least the eastern half of North America (they reached into Nevada and Washington as well) a century earlier. The ornithologist W. L. Dawson, in his 1903 treatise *The Birds of Ohio,* wrote, "No more marvelous tales have been handed down to us ... than those that our own fathers tell and solemnly asseverate, concerning the former abundance of the Wild Pigeon during their migrations and in their breeding haunts. During their passage, the sun was darkened and the moon refused to give her light. The beating of their wings was like the voice of thunder and their steady on-coming like the continuous roar of Niagara."[8]

Passenger pigeons were good to eat too, a staple on nineteenth-century tables, as this recipe from an 1808 cookbook attests (I've preserved the typography of the original, in which ſ substitutes for *s*):

> Lay a thin ſheet of paſte round the rim and ſides of a deep diſhe, ſprinkle a little pepper and ſalt on the bottom, and put in a thin beef-ſteak; pick, draw and ſinge ſix pigeons, waſh them clean, cut off the feet, and ſtick the legs into the ſides, ſeaſon the inſides with pepper and ſalt, put a little butter in the inſide of every one, put them in the diſh breaſt upwards, and the neck ends next to the rim of the diſh; put the gizzards between them; ſprinkle

some pepper and ſalt over them, and put in a gill of water; lay a very thin ſheet of paſte before it is puft over them; and with a bruſh wet the paſte all over; then put a ſheet of puff-paſte half an inch thick over that, cloſe it, rub it over with the yolk of an egg, ornament the top, ſtick the feet in, and bake it nicely; when it is taken out put in ſome good gravy, and ſend it to table hot. You may put in the yolks of ſix hard eggs, or leave out the beef-ſteak, if you think proper.[9]

When passenger pigeons flocked, there were so many of them roosting on trees that their weight broke the limbs, as described again by Dawson: "Even trees two feet in diameter, were broken down beneath their weight, and where they nested a hundred square miles of timber groaned with the weight of their nests or lay buried in ordure."[10] Buried in ordure is not a pretty picture—ordure is dung.

Given their dense numbers, it was easy to cast nets over the pigeons while they roosted or to blast bunches of them out of the sky with a single shotgun shell. But by 1878 passenger pigeons were going fast. Dawson reported the take from one nesting field near Petoskey, Michigan: "The number of dead birds sent by rail was estimated at 12,500 daily, or 1,500,000 for the summer … an equal number was sent by water. '… adding the thousands of dead and wounded ones not secured, and the myriads of squabs left dead in the nest, at the lowest possible estimate, a grand total of 1,000,000,000 pigeons sacrificed to Mammon during the nesting of 1878.' "[11] Even if this number is a hundredfold exaggeration, as Dawson believes it was, losing a million birds per year in a single nesting area, multiplied by many nesting areas, could not be sustained for too many years. And of course, it wasn't, which is why this is what Dawson listed for the known range of the species in 1903: "Breeding haunts unknown. Formerly exceedingly abundant migrant and summer resident. Bred locally in vast numbers, now almost unknown. 'Last records' are coming in from various quarters, but they are mainly ten to twenty years old."[12]

Luckily for those pushing to re-create passenger pigeons, the skin and feathers of Martha and several of her kin have been safely tucked away for posterity in museum drawers. Those hold the raw materials the de-extinction crew needs, namely passenger pigeon DNA. And here's where things get tricky. As I mentioned earlier, DNA degrades fast, so even in well-preserved skin and feathers, only bits and pieces of what used to be a complete genome are intact. That means that the simple version of cloning is out of the question—that takes a whole viable genome, and even when you have that, it generally takes many unsuccessful attempts to produce a single successful result, and most animal clones do not live very long.[13]

So you have to pull some molecular tricks. The first trick is to replicate as many as possible of those bits and pieces of ancient DNA (so called because the DNA is extracted from preserved museum or fossil specimens) from passenger pigeons, arrange the pieces in the right order, figure out what bits are missing, and then somehow fill in the missing pieces. That is no easy task, because when you start, you don't actually know what the right order is, or which pieces are gone. To figure that out, you use the genome of the nearest known living relative of the passenger pigeons—the band-tailed pigeon (*Patagioenas fasciata*)—as a template to help order the pieces of passenger pigeon DNA that you were able to recover. That requires, first, sequencing the entire band-tailed pigeon genome, which has not yet been done but is quite feasible with modern genomic techniques. Then, with enough work, you just might be able to take the isolated segments of ancient DNA you've recovered from the passenger pigeon museum specimens, arrange them in order by comparing them, base pair by base pair, with the complete band-tailed pigeon genome, and through that process figure out what parts of the passenger pigeon genome you are missing. Once you know that, you fill in the blanks by substituting what seem to be corresponding DNA segments from band-tailed pigeons. What you end up with is a sequence of base pairs—strands of DNA that when hooked together properly compose the genome—that is, in reality, part passenger pigeon and part band-tailed pigeon. That may be ok—we share 99 percent of our DNA with chimpanzees, and you would expect closely related pigeon species to be mostly genetically identical.

Then the going gets really tough, because you have to make a germ cell that carries viable copies of the reconstructed passenger pigeon genome. Germ cells are the ones that develop into sperm or eggs. Given that there are no germ cells in dead passenger pigeons, you have to make them from scratch. The first step is to make a whole genome that resembles a passenger pigeon's. Remember, at this point, you still don't have the actual, physical genome of the extinct bird. Instead, what you have is the knowledge of what order the base pairs should be arranged in. To make the actual genome, you have to genetically edit parts of the viable, fresh DNA of a band-tailed pigeon, base pair by base pair, such that it takes on the same sequence of base pairs as a passenger pigeon's DNA.

This is no easy feat, for a couple of reasons. First, genetic editing on the required scale is by no means routine—so far, it has been accomplished only for fine-scale editing of bacterial genomes. Second, there are likely to be millions of differences in the two genomes, given the thirty million years of evolution that are thought to separate the two species. Since it is impractical to edit every base pair to exactly replicate a passenger pigeon genome, you have to figure out which of the millions of differences are the significant ones for turning a band-tailed pigeon into passenger pigeon. That is not trivial; we don't even know the function of the vast majority of human genes, let alone the genes of species like band-tailed pigeons, which haven't even been completely sequenced yet.

Assuming the band-tailed pigeon's DNA gets edited to appropriately mimic passenger pigeon DNA, the next steps are to insert it into the germ cell of yet a third species of pigeon, perhaps a rock pigeon (*Columba livia*), which is easy to work with in the lab, and hope that the modified germ cell ends up in the gonads of a developing pigeon chick. That chick would still be a rock pigeon, but it would carry viable germ cells of the passenger pigeon mimic. This step is clearly just on the drawing boards at this stage—it's complicated by the fact that developmental biologists still don't know how to extract the needed germ cells from pigeons, though it has been accomplished for chickens, and perhaps burgeoning stem cell technology will come to the rescue. The final step would require breeding two of the rock pigeons that carried passenger pigeon germ cells, which, in theory, should result in the birth of a passenger pigeon—or at least something like a passenger pigeon.[14]

All this would require the investment of millions of dollars, and if everything proceeded smoothly, at least ten years of work.[15] That means the soonest the first re-created passenger pigeon would peck its way out of its eggshell would be around 2023. At that point, could we really claim de-extinction of passenger pigeons? In fact, what we would have created is a species made out of leftover passenger pigeon parts spliced together with spare band-tailed pigeon parts. Albeit in a much more sophisticated and feasible way, we would have achieved what Mary Shelley's Dr. Frankenstein attempted to do with the leftover parts of dead people. The outcome of Dr. Frankenstein's experiment, no surprise, was a misunderstood, very unhappy monster unable to survive in the world into which he was thrust.

That may be an overly dramatic comparison to make with attempts to re-create a passenger pigeon, but in the context of what that genetically engineered pigeon would do for bringing back an entire species in all its glory, it may not be too far off. Almost certainly, the first passenger pigeons so produced would be oddities, raised by people in controlled zoo environments—given the millions of dollars it would have taken to produce them, they'd be too valuable to set free to see if they could make it in the real world. If their numbers grew to the point where releasing them into the wild became feasible, as is the eventual goal, there would be no passenger pigeon parents to teach them how to be real passenger pigeons. That knowledge died forever with Martha. The suggestion that we could use homing pigeons to train them to be passenger pigeons is, frankly, a little ridiculous.[16] That's about the same as asking a Chihuahua to teach a wolf pup how to thrive in the wild.

Even if the re-created passenger pigeons were successfully reintroduced into their forebears' home range, the ecological fabric that extinct passenger pigeons were part of no longer exists. The chestnut forests they frequented have all been cut down or died from an invasive blight, and the vast majority of Martha's forebears probably never saw a European starling (*Sturnus vulgaris*). Now there are about 200 million starlings living in what used to be passenger pigeon territory.[17]

My point is that even after the decade of work and millions of dollars, we still wouldn't have replicated what extinction has taken. And passenger pigeons are the best-case scenario: they haven't been extinct too long; there are lots of specimens in museums, so it might be possible to create a genetically variable population; there are close living relatives to use as a genetic template and germ cell incubator; and their generation times are relatively short. Once we begin to move toward species like mammoths, the problems multiply considerably. For mammoths, highly fragmented DNA sequences, lack of any living relative closer than the Asian elephant, and—judging from living elephants—a two-year gestation time and about twelve years to reach sexual maturity mean that the most optimistic timeline for producing some sort of viable mammoth-elephant hybrid would be when my kids have kids, or maybe grandkids.[18]

Then there's the obvious problem of mammoths being ice-age creatures—they thrived when glacial ages prevailed, and their populations shrank during warm interglacial times.[19] In fact, it seems that a combination of warming climate and human pressures caused them to go extinct.[20] Reconstituted mammoth-elephants would be born into a time hotter than any mammoth ever lived through, even if we produced them today, and as we as we learned in chapter 4, it's only going to get hotter in the coming decades. That makes the loss of chestnut forests for passenger pigeons seem trivial. And of course, if the demand for ivory keeps up, what makes anyone think that the new mammoth ivory wouldn't be even more valuable than elephant ivory? Which would mean the reconstituted mammoth-elephants would have only one relatively safe place to live—in a fortified zoo.

So today the answer that I would give a student who asked me about technology bailing us out of the extinction crisis would be slightly different from the answer I gave in 2012, a bit more measured, a bit more detailed. It would boil down to this, though: the most likely contribution that molecular biology will make to conserving biodiversity is a few individual genetically engineered animals that live in zoos, primarily hybrids that mix the traits of living species and extinct ones, second-best replicas of what used to be. At the end of the day, that doesn't do much to solve the Sixth Mass Extinction crisis, because it fails to address the real root of the problem, which is how fast still-surviving species are disappearing.

That, in fact, is exactly what has many conservation biologists dead set against de-extinction. Many of them see it as standing in the way of preventing extinctions by giving us false hope that technology will do the hard work for us and by diverting financial resources from on-the-ground efforts to stop the last populations of various threatened species from dying out. Stuart Pimm, one of the world's leading conservation biologists, summarized the argument against de-extinction this way: "Conservation is about the ecosystems that species define and on which they depend. Conservation is about finding alternative, sustainable futures for peoples, for forests, and for wetlands. Molecular gimmickry simply does not address these core problems. At worst, it seduces granting agencies and university deans into thinking they are saving the world. It gives unscrupulous developers a veil to hide their rapaciousness, with promises to fix things later. It distracts us from guaranteeing our planet's biodiversity for future generations."[21]

A few, myself included, would also point out an irony about the de-extinction effort: for all the forward-looking technology that de-extinction emphasizes, it falls into the same trap that conservation biology fell into initially. It focuses on re-creating the past, rather than keeping nature evolving into the future. In recent years, the science of conservation biology itself has evolved, away from that focus on the past and toward viewing nature for what it is: an ever-evolving dynamic in which what you knew as "normal" as a nine-year-old is no longer the norm when your kids grow up, just as what your parents knew when they were kids was no longer "normal" by the time you grew up.[22]

•

That is different, in a not-so-subtle way, from what has been the implicit mantra of conservation science for at least the past half century (and that currently afflicts the de-extinction movement), which has basically been this: "to preserve, or where necessary to recreate, the ecologic scene as viewed by the first European visitors."[23] Those were the marching orders laid out in a landmark report, commissioned by the United States National Academy of Sciences and published in 1963, on how to manage national parks in the United States. The study became known as the Leopold Report because it was authored by a committee headed by A. Starker Leopold, the son of conservation pioneer Aldo Leopold. You can substitute whatever time frame and culture you want for "as viewed by the first European visitors"; the operative words are "preserve or recreate."

But of course, nature isn't static, so trying to hold it to a fixed point becomes problematic, from both a philosophical and a practical perspective. By 1985, the mission of conservation biology was more broadly defined as being to "advance the science and practice of conserving the Earth's biological diversity."[24] Still, there remained the underlying influence so clearly articulated in the Leopold Report, that the conditions our grandfathers experienced, or the ones we knew as kids, were "right" for nature, and that it was important to keep it from straying too far from that.

The folly of striving to keep nature static is apparent if you look at how much things change from century to century, millennium to millennium, or over even longer time scales. Over the shorter term, however, say, year to year or decade to decade, which are the time frames people tend to recognize as important, the goal of holding nature constant seemed to be working pretty well for conservation biology up through the last half of the twentieth century. That's because, usually, the time scale over which ecosystems change is slow compared to the lifetime of a person. When I was a kid, for example, my dad took our family camping in some of the same parts of the Colorado Rockies that he had hunted and fished as a teenager. The elk were still there, the fish were still in the streams, the same trees were in the forest, and that had been the case since my dad was a little kid, even longer. In growing up, he had changed much more than that ecosystem had.

That's what's different today. When I go back to those places where I camped with my dad, and later, on my own as a teenager and young man, I find that the ecosystem has changed much more than I have. I have to drive farther to get away from the houses and clear-cuts. Once I'm there, most of the forest is dead from climate-triggered beetle kill or from drought-amplified forest fires, and the snow in the high country is gone, even though it's just barely early summer. Some of the animal species are different—I'm apt to see a robin in the middle of winter now, even though they used to be harbingers of spring.

That's not confined to where I grew up—it's happening worldwide. Over the same time period, we began to see changes even in highly protected and very remote ecosystems. For example, in Yellowstone National Park, the world's oldest, where nature has been protected since 1876 across an area that spans parts of the states of Wyoming, Montana, and Idaho, amphibians were declining drastically by the early twenty-first century. Remote lakes in the far north of Canada and Alaska, far from any direct influence by humans, show signs of accumulating unusually high levels of nitrogen, reflecting global changes in biogeochemical cycles.

In short, ecological change worldwide has sped up so much that it now outpaces, and is observable within, a single human lifetime. Under those circumstances, it's become obvious that trying to hold nature to the condition it was even in 1950, let alone when Europeans first saw a place, is a lot like spitting into the wind. In view of that stark reality, some conservation biologists have begun to change their tune. Instead of regarding nature as something to be preserved and kept separate from humans, a vociferous faction of conservationists has begun to espouse the philosophy that people are part of nature and always have been (well, at least since we evolved), that nature itself is dynamic, and that trying to hold the line against change could in the long run be counterproductive to conserving species and biodiversity. Instead, they argue, the most effective way to save species may be to embrace change and try to guide it in directions that best balance the needs of people with those of the ecosystems they depend upon.

That goes against the grain of what used to be the tried-and-true approach to conservation biology, the approach that actually worked pretty well for most of the twentieth century. That is kind of a one-size-fits-all model that considers humans as largely separate from nature; it regards keeping people out as the best way to protect everything that needs protecting. In that model, by setting aside a large enough tract of land (or, less commonly, sea) and minimizing direct human impacts, you automatically preserve, in one fell swoop, all of the things considered important in the conservation community—individual species and populations, whole ecosystems and ecosystem services, and the wildness of wild places. Over the past few decades, that has resulted in some very good things for nonhuman species and humans alike, not the least of which is that 12 percent of Earth's lands are now protected in some sort of nature reserve, which has kept many species alive.[25] Undeniably, the strategy of protecting those places has been absolutely crucial to keeping as many species on Earth as we still have. There would be, for example, no grizzly bears left in the lower forty-eight United States if it weren't for the safe haven afforded them by Yellowstone and Glacier National Parks. And it's likely that rhinos would have already bitten the dust in Africa without the proven approaches of restoring and protecting habitat and safeguarding the rhinos themselves.

But conservation biologists themselves are quick to point out that this "protect and preserve" approach by itself hasn't been enough. For instance, while conservation efforts from 1996 to 2008 prevented at least twenty-eight mammal species from proceeding farther along the extinction trajectory, over the same time, 156 mammal species slipped one IUCN category (for example, from vulnerable to endangered) closer to their demise.[26] When you throw into the mix things like climate change, the extent of habitat fragmentation we now have, poaching financed by international crime rings and terrorist organizations, and the addition of a couple billion people over the next three decades, it becomes clear that limiting ourselves to setting aside nature preserves will no longer work. For one thing, there are just not that many places left to set aside. And what do we do when, for example, the climate changes such that a protected area can no longer support the species it was set aside to protect, as is anticipated for places like Joshua Tree National Park in California?[27]

In the new—and I might add, still very controversial—view of conservation biology, a big part of the answer is to shed the concept that nature is what people set aside to keep wild, and to replace it with the concept that nature is an ever-changing spectrum we live within and have already modified far beyond any preconceived notions of "wild." In the words of some of the strongest proponents of the new conservation biology, this means a new mandate for the science:

> Instead of scolding capitalism, conservationists should partner with corporations in a science-based effort to integrate the value of nature's benefits into their operations and cultures. Instead of pursuing the protection of biodiversity for biodiversity's sake, a new conservation should seek to enhance those natural systems that benefit the widest number of people, especially the poor. Instead of trying to restore remote iconic landscapes to pre-European conditions, conservation will measure its achievement in large part by its relevance to people, including city dwellers. Nature could be a garden—not a carefully manicured and rigid one, but a tangle of species and wildness amidst lands used for food production, mineral extraction, and urban life.[28]

In fact, Emma Marris argued cogently in her recent book that the world is now just such a garden—a rambunctious garden, she called it—that has been almost entirely shaped by our ubiquitous influence on the planet.[29]

This view of conservation biology does not go over well with many in the field, and indeed, some of the arguments for the "new" conservation biology are more rhetorical and exaggerated than helpful.[30] For instance, who cares whether or not Edward Abbey "pined for companionship" during his forays into the Utah deserts or whether he always practiced what he preached; that doesn't mean that there is no solace to be found in the places he wrote about.[31] I know; I've been to many of those places, and it's pretty much like Abbey said. What produces the solace is precisely the fact that humans seldom visit there and wide-open landscapes and wild species predominate. Rhetoric aside, though, (and to be fair, rhetoric and exaggeration happen on both sides of the argument), the newly emerging view of conservation biology brings in perspectives that will be helpful in developing strategies to save species in the future. The key perspective is that it's essential to integrate other species into human-dominated systems—because that now describes the condition of most of the planet. It describes the whole planet when you factor in indirect human impacts, like climate change.

We are, in fact, already conserving many species in ecosystems we dominate, in ways that have helped keep their populations viable. On the wilder side, populations of wolves and mustangs are maintained alongside beef cattle in the western United States (not without strong opposition by ranchers, however).[32] Rhino farms, in addition to the traditional approach of protecting rhinos in nature preserves, have been helpful in rebuilding southern white rhino numbers after the near annihilation of the species. Mountain gorillas survive in Rwanda, Uganda, and the Democratic Republic of the Congo only by virtue of the tourist dollars they bring to the local economies. And zoos, which play a crucial role in preventing extinction of highly endangered species, are popular worldwide.

None of these examples is news, of course, to "traditional" conservation biologists—quite the contrary. These are approaches that those practicing in the field for decades developed. What sets the "new" conservation biology apart is the underlying philosophy that if maximizing biodiversity is the goal, *all species and landscapes* must be managed heavily; that inevitably, we have to throw in the towel on keeping wild places wild. It actually goes further than that: the tenet is that there's *no* "wild" left, anywhere. That is a hard pill to swallow for those of us who view the wild side of nature as its best side, and it brings up deep questions. For example, is it worth it to keep, say, lions or tigers or grizzly bears in zoos if they can live nowhere in the wild? I might actually answer that question with a "no," though I'm not a hundred percent sure. For me, the value of those species is that they signal I'm in a place where I don't rule, where I have to think hard about what my life means in the bigger scheme of things, and where, in order to avoid getting eaten, I have to wake up some senses that are usually deadened. I don't dwell on why I like the feelings I get in that kind of nature, I just know that, like tens of millions of others, I do like those feelings, and my life would be poorer without them.[33]

On the other hand, I also know plenty of people who feel a lot better in their backyard than they do in a place that can only be reached by walking or riding a horse. I know others who would say, without hesitation, if the only place tigers can survive is in a zoo, let's keep them in zoos, the world is still better off with them than without. Still other people have said to me—this from another student in another freshman seminar—"If it comes down to between keeping elephants alive and people alive, elephants gotta go."

The point is that people express many different reasons for wanting to keep species alive: the moral imperative, liking certain species, banking diversity for the future, ensuring ecosystem services, preserving certain versions of nature, cultural tradition—the list goes on. The most effective way to pay off the extinction debts that have accumulated over the past few centuries, and to prevent more from stacking up, is to take advantage of every one of those reasons.[34] But don't fool yourself, or try to fool anyone else, into thinking your personal favorite addresses every reason people have for saving species. That's the slippery slope to bickering about whose view is best, which in the end is counterproductive. This is what seems to be happening with the polarization between de-extinctionists, old-school conservation biologists, and new-school conservation biologists.

De-extinction, for example, may well help uncover ways to increase genetic diversity, but it will not help repopulate the missing parts of ecosystems. If you're a rhino farmer, flooding the Chinese traditional medicine market with southern white rhino horns might keep that subspecies alive, but it won't do much for the four out of five species not farmed—they'll still be free money for the poachers. Fixing the problem for those poached species that can't be farmed will take some tough guards with big guns. Pushing ecosystem services is great for conserving the little-recognized species that provide obvious benefits like water filtration, pharmaceuticals, and ecotourism, and it helps focus the public on the value of nature, but it will not do much to slow the loss of snow leopards in the Himalayas or Yellowstone wolves that cross over onto a rancher's land. National parks and wilderness areas will provide refuges for species that need large tracts of land to roam and will preserve a feeling of wildness, but they won't hold nature constant given global pressures like climate change—some species that live in a given place now will disappear, and others will move in. In fact, to maximize species survival, even those wild-feeling places may end up being managed more than we're used to, when it becomes necessary to help species skip over fragmented landscapes in order to get from a place where their needed climate is disappearing to a place where they can still survive.[35]

This last point exemplifies that even though there are now many arrows in the conservation quiver, we are going to have to add new ones in coming years. It will, for example, be necessary for governments at all levels to provide incentives to save species and to devise and enforce strict laws that discourage extinction-causing practices. New marine reserves (yes, that old-school approach) will be required if we have any hope of saving many threatened species in the sea. For existing protected lands, the mandates may need to be revised to specify regions whose sole purpose is preservation of species, versus areas whose chief function is allowing nature to evolve with as little interference as possible.[36] In the former, transplanting species to save them from climate change or trucking in water to maintain watering holes might be advisable. In the latter, those activities would be strictly prohibited, even if it meant some species died out there. The goal in those places would be to let nature take its course without our constant attention, which in the future may be as close as we'll be able to get to having truly wild places. In addition, there may be a role for brand-new kinds of species reserves designed to mimic the past—museums of nature, if you will. These have already been proposed by some conservation biologists under the rubric of "Pleistocene rewilding," which would repopulate places like the American Midwest with ecological analogs of the megafauna that once were native there.[37] For example, to substitute for mammoths, elephant populations would be established, and Old World lions and cheetahs would fill in for extinct saber-toothed cats and so-called North American cheetahs. If at face value this sounds absurd, it's not so different from the game parks that already exist in places like Texas or South Africa. Make no mistake: rewilding would not bring back Pleistocene ecosystems, because not only the Pleistocene animals but also the Pleistocene climate and vegetation assemblages are gone forever. Nor would they be wild ecosystems; they'd be the antithesis of wild, because we would have built them. They would, however, enhance the survival of many threatened large mammals by increasing their geographic distribution, and maybe, just maybe, when we visited those places they would thrill us and reawaken some feelings that had lain dormant. Of course, nobody is crazy about having animals like elephants, lions, and cheetahs tromp through their farmland, even in Africa, where they are native. So ironically, yet another effort toward keeping the wild species of the future wild likely will involve miles and miles of fences.[38]

I have to admit that thinking of those fences, and all the other ways that nature will change if we are to save the maximum number of species in the coming years, makes me a little sad. I'd love to freeze in time the way the world seemed the morning I was walking across the rolling hills of southeastern Kenya with my wife, two friends, and two Maasai warriors. There were no fences, but there were elephant tracks and warm sun, and as we walked, one of our Maasai companions, dressed in his traditional red tunic, explained to me why having wild animals around is so critical for maintaining his traditional culture. He also explained how his culture was changing, how the modern world was moving in fast.

That fast-changing world, of course, is the real world I was walking through that morning, and it is the world we all live in today. The other reality is that we're the agents of change, especially when it comes to deciding whether to watch species go extinct or to nurse them back to the point where they can make it on their own. No doubt, going full steam ahead with the multipronged, heavy-duty management required to resuscitate species we've brought to the brink will fundamentally change humanity's relationship both to those species and to nature itself. But the alternative—watching species rapidly disappear, day after day—means that we sever those relationships altogether.

CHAPTER EIGHT

Back from the Brink

Here's the catch: what's bad for other species is also bad for us. Just as they cause problems for other species, our current methods of pursuing power, food, and money no longer work very well for sustaining the human race. Increasing numbers of studies, including one signed by thousands of scientists from all over the world, come to conclusions like "Based on the best scientific information available, human quality of life will suffer substantial degradation by the year 2050 if we continue on our current path."[1] The drivers of demise that are invariably cited are, you guessed it, the once effective but now outdated and harmful ways we choose to generate the power, food, and money that seven billion people currently rely on and that will be needed by the additional two and a half billion who will join us by 2050.

The year 2050 is not that far off, as my kids point out to me. They won't even be as old then as I am now. But 2050 does seem far off to some, so here's another perspective: you don't have to wait until 2050 to see how people are being hurt. All you have to do is read the newspaper. For instance:

> **European Coal Pollution Causes 22,300 Premature Deaths a Year, Study Shows** … A total of 240,000 years of life were said to be lost in Europe in 2010 … [2]

> **Corpses Rot, Flow down the Ganga in Uttarakhand** ... The estimated death toll [from the floods] could be as high as 15,000 ... [3]
>
> **Two Killed as Colorado Wildfires Destroy 360 Homes, Force Evacuations in Colorado Springs** ... 38,000 people in the region were forced to flee ... [4]
>
> **Texas Gov. Rick Perry Extends Drought Emergency in Most of State** ... Low rainfall and record high temperatures have reduced water supplies and aquifer levels, threatening many parts of the state ... [5]

These headlines give just a taste of the mounting costs in human lives and livelihoods lost as a by-product of generating most of our power from fossil fuels. Air pollution, coming largely from coal plants and vehicle exhausts, is a direct killer, causing six million deaths a year worldwide.[6] No wonder, considering you can even see it from space on bad days.

Also bad news for people is the increased frequency of extreme weather events that result from the way fossil fuel emissions are changing climate. Extreme weather events are catastrophes like droughts, heat waves, intense flooding rains, and so on. People have always experienced these natural calamities, but heating up the atmosphere by injecting greenhouse gases into it causes extreme weather events to become both more frequent and more powerful, for reasons I mentioned in chapter 3.

If you doubt that that is already happening, look at the statistics on the frequency of annual weather catastrophes that carry at least a billion-dollar price tag for taxpayers in the United States. The number of those high-cost events has increased about 5 percent per year since 1980.[7] In the five years spanning 2008 to 2012, there were forty-four such calamities in the United States, about twice as many as in any previous ten-year period.[8] Twenty-five of the billion-dollar catastrophes hit in just two recent years (fourteen in 2011 and eleven in 2012). In each of those years, the number of such events was higher than had ever occurred in a single previous year.[9] In 2012, the climate catastrophes included multiple tornadoes in the Southeast, hurricanes Isaac and Sandy, drought-fueled western wildfires, and a severe heat wave that dried up crops across the North American continent. The first half of 2013 saw western wildfires and tornadoes more devastating than those of 2012, with floods in the Midwest thrown in for good measure. And as I write this in early 2014, California is experiencing its driest year in history.

It's the same story worldwide: in the first half of 2013, the worst floods in centuries hit Europe and parts of Asia, and drought devastated Australia and East Africa. These disasters come on the heels of other extreme weather events that have been building in frequency and intensity over the past decade. In western Europe, dangerous heat waves are now twice as frequent as they were in the previous century. Some 40,000 people died from the heat that cooked western Europe in August 2003, the hottest summer there in 500 years. In 2010, a Russian heat wave killed 56,000 people.[10]

That's not as bad as it's expected to get if we keep on doing business as usual. In that eventuality, by the end of the twenty-first century, the scorching heat waves that now occur once every twenty years are expected to occur every other year in most regions.[11] The worst-case projections suggest that if the current trajectory of warming continues to the year 2100, some areas where people now live will simply be too hot for humans to survive.[12]

Similarly, in many regions where it has been normal for flood-causing deluges to occur once every twenty years, the deluges are forecast to occur as frequently as every five years.[13] This means that storms like the one that caused water to fill the New York City subway system in 2012, or that put nearly one-fifth of Pakistan under water, killing 2,000 people and displacing 18 million more in 2010, would happen much more often.[14] Sea level rise from melting glaciers and thermal expansion (sea water expands as it gets warmer) is expected to displace more than 100 million people. In regions that depend on seasonal snowpack or glaciers and a gradual spring melt to supply water for crops and for major cities, less snow and rapid melting are expected to cause major water shortages for a billion urban dwellers. Afflicted cities will include Los Angeles, Mumbai, Delhi, Beijing, Mexico City, Lagos, and Tehran.[15]

The story for food is much the same. Doing business as usual is beginning to show its limits. Again, recent headlines pretty much say it all:

> **Uganda: Severe Food Shortage Hits Karamoja** ... Death has occurred in Kaabong as a result of famine ... [16]
>
> **UK "A Few Days Away" from Food Shortage** ... Retailers and food producers should be penalised for wasting food, according to a hard-hitting new report ... [17]
>
> **UN Warns of Looming Worldwide Food Crisis in 2013** ... We've not been producing as much as we are consuming ... [18]
>
> **Legacy Food Production Techniques Won't Feed Population of 2050** ... Clearly, the world faces a looming agricultural crisis ... [19]

The stories associated with these headlines highlight what's wrong with our current system of feeding the world. First, food production in those countries where the population is growing most rapidly is insufficient to feed local people. In those food-insecure places, hunger increased from 1990 to 2011, for example in Africa, where the number of malnourished children went from 46 million to 56 million in those twenty years.[20]

Second, where there are food surpluses, distribution mechanisms have never been developed to the point where the surplus can be delivered to the places that need it most. Instead, the excess food simply gets thrown away; as mentioned in chapter 5, over a third of what we grow is wasted. The double whammy of not growing enough food locally and not being able to ship it from areas of surplus contributes mightily to one out of every eight people on Earth—nearly a billion of us—being chronically hungry.[21]

Third, dependency on a worldwide food trade is becoming unsettling even to countries like the United Kingdom, where food supplies are generally considered adequate. The fear is that worldwide shortages could cause sudden price shocks that would result in supermarket shelves emptying within days.

Fourth, those fears are well founded, judging from analyses that look at both the past and the future of food. Tracking past trends reveals that food consumption has exceeded the amount grown for six of the past eleven years. That has drained many countries' food reserves. Reserves have gone from an average of 107 days' worth of food ten years ago to less than a 74-day supply.[22] The future looks even less rosy unless we change the way we presently grow, distribute, and waste our food. A recent study by food-security researchers at the University of Minnesota's Institute for the Environment pointed out that continuing business as usual will require us to double our crop production by 2050, which means increasing production by 2.4 percent per year.[23]

We haven't been doing that. Production of four principal crops—maize, rice, wheat, and soybeans—has only been increasing by 1.6, 1.0, 0.9, and 1.3 percent respectively since 1989. Continuing at that rate will mean many more hungry people by 2050. These projections do not take into account disruptions in the food supply that are likely to result from climate change—disruptions triggered both by extreme weather events, like the 2012–2013 drought that decimated the United States corn crop, and by general warming, which reduces crop yields due to more frequent high-heat days, even when moisture is adequate.[24]

The hunger problem has another source, of course, which is the third driver of the extinction crisis: money. Eighty percent of the world's population—5.6 billion people—have a tough time buying food, even if it is available, because they live below the poverty level, defined as subsisting on less than $10 per day. Of those, 1.4 billion subsist on less than $1.25 per day.[25] Poverty-related hunger is not restricted to developing countries—in the United States in 2011, about 50 million people lived in food-insecure households, about the same number as those who fell below the U.S. poverty threshold.[26]

The optimistic projection is that poverty will end by 2030, and indeed, the trend over past decades has been heading in that direction if you define poverty strictly enough.[27] However, that optimistic projection refers only to bringing those people making less than $1.25 per day over that threshold. I'm guessing that those of you reading this book would not be too satisfied if your daily spending limit was $1.26, and I doubt that those people whose income rises to, say, $2.00 per day are going to consider themselves a lot better off.

In fact, other views of how the world economy is changing indicate we are going in the opposite direction. One concern is the growing proportion of the world's poor in middle-income countries like Brazil, Argentina, Chile, South Africa, China, and the Russian Federation, where per capita gross national income is between $1,026 and $12,475 per year.[28] From 2005 to 2010, the percentage of the poor found in middle-income countries increased from 26.5 percent to 65.9 percent.[29] In part, this is because twenty-six countries "graduated" from low-income to middle-income classification in the past decade, including India, Nigeria, Pakistan, and China. But the shift in demographics also reflects that as economies in low- and middle-income countries grow, inequality between the richest and the poorest and between households, governments, and corporations increases.[30]

This does not promote stable political situations, to say the least, which may contribute to the apparent coincidence between growing numbers of failed and fragile nations and the growing proportions of the poor in those nations. These are countries that cannot assure "basic security, maintaining rule of law and justice, or providing basic services and economic opportunities for their citizens."[31] Examples include Nepal, Bosnia and Herzegovina, the West Bank and Gaza, the Democratic Republic of the Congo, Rwanda, Somalia, Afghanistan, and the Federated States of Micronesia—forty-seven countries as of 2013.[32] The number of failed and fragile states has risen from twenty-eight in 2006 to forty-seven today, indicating how rapidly political instability is spreading, at the same time that the percentage of the world's poor in such nations has doubled from 20.5 percent to 40.8 percent.

The past couple of decades have also seen a troubling global trend toward greater inequity between rich and poor in the richest countries, as measured by the Gini coefficient. A Gini coefficient of zero means everybody has the same income, whereas a coefficient of one means that a single person or income group makes all the money and everybody else is broke. In the United States, for example, the Gini coefficient went from 0.39 in 1968 to 0.47 in 2010, which translates to a 20 percent increase in the gap between rich and poor. Of the thirty-four countries in the Organisation for Economic Co-operation and Development (OECD), only Turkey and Mexico have more unequal societies than the United States.[33]

The United States is not alone in its trend toward less income equability. Between 1980 and 2010, inequality between rich and poor increased in seventeen of the twenty-two OECD countries for which data were available. Included in that list are Mexico, the United States, Israel, the United Kingdom, Italy, Australia, New Zealand, Japan, Canada, Germany, the Netherlands, Luxembourg, Finland, Sweden, the Czech Republic, Norway, and Denmark. The list gets longer if you look at only the past decade, a period for which data are available for more countries.[34] Increasing income inequity is bad for humanity as a whole for a host of reasons, most of them basic to a healthy, well-functioning society. For instance, increasing income inequity correlates with less well-being in children, higher dropout rates in schools, more violent crime, more mental illness, less trust among people, and less opportunity for social mobility.[35]

Having most of the money flow through just a few hands also tends to make for unstable world economies, as has been shown by analyses of financial transactions that led up to the economic crash of 2008.[36] In 2003, the transactions of financial investment institutions, real estate companies, technology companies, oil companies, and other basic material companies were largely separate from each other. A network diagram that shows how daily market returns were connected among those corporations looks like a Rorschach inkblot that I'd call a lobster. The tail is a cluster of technology corporations, only very weakly connected to the lobster body of investment institutions. Off to the upper right, held on to the "body" by only a few strands of interactions, are the real estate companies. The other "claw," also only weakly connected to the "investments body," is the cluster of oil companies and other basic materials providers. As you progress through 2004, 2005, 2006, and 2007, the blots progressively morph; the tail and claws become much more firmly joined to the body and finally just melt into it. By 2008—when the world's economy plummeted into the Great Recession that we're still digging out of—the network diagram looks like a big blob with no definable appendages. What happened over those six years is that more and more transactions began to flow through the same financial investment companies rather than being segregated in several different sectors of the economy, setting up a major instability. When the financial firms crashed, so did everything else.

That is where doing business as usual has led us: millions of people suffering and dying from climate change and fossil fuel pollution; food shortages that promise to get worse; and an economic system that not only crashed very recently but is getting more and more unstable as a result of ever-growing inequity. How do we make sure the bad things we already see happening do not get worse? By implementing the same solutions that I laid out earlier for dodging the Sixth Mass Extinction. Just as what's bad for other species is bad for us, it turns out that what's good for other species is also good for us.

Pulling back from the brink of disaster will not be easy, of course, given that it will take changing the way we're used to doing some basic things—producing energy, feeding the world, making money. Changing each of those is a very, very big task. So big, in fact, that the initial reaction of people is simply to shake their heads, throw up their hands, and say, "It can never happen." That, however, is where you'd be wrong, as past experience shows.

Solving problems on this gigantic scale requires that a few things come together. The first is simply recognizing that a global problem exists and that not fixing it will be a bad thing for humanity. The second is understanding that fixing global problems requires win-win interactions between local communities—where changes actually take place through local people's efforts—and governments, which have to define priorities and back them with clear incentives. And of course, cooperation across local, state, and national boundaries is essential, since solving global problems ultimately requires global vision. The actual on-the-ground solutions demand individual initiative, technological advances, and emplacement of new infrastructure, which allow us to do things in new ways. As the 2013 statement "Scientific Consensus on Maintaining Humanity's Life Support Systems in the 21st Century" puts it: "Individual initiative has seldom been in short supply and continues to be a powerful human resource. Successful global-through-local cooperation resulted in ending World War II and rebuilding afterwards; banning use of nuclear weapons; dramatically increasing global food production with the Green Revolution and averting food crises through United Nations initiatives; greatly reducing the use of persistent toxic chemicals like DDT; reversing stratospheric ozone depletion (the 'ozone hole'); and diminishing infectious diseases such as malaria and polio worldwide."[37]

The third thing it's important to know about solving global problems is that it doesn't happen in a heartbeat. You can't turn a giant ocean liner on a dime, as the captain of the *Titanic* learned firsthand. Likewise, you can't change the trajectory of things as big as energy production, the food industry, and investment strategies overnight. But once people decide something needs fixing, it is remarkable how quickly we turn thought into action and cause incredible advancements to take place, as noted in chapter 4.[38] More examples are not hard to find; again from the 2013 scientific consensus statement: "Past technological advances and the building of new infrastructure have been remarkable and commensurate in scale with what is needed to fix today's problems. For instance, ... [in less than fifty years] 60% of the world's largest rivers were re-plumbed with dams. In about 30 years, the world went from typewriters and postage stamps to hand-held computers and the internet, now linking a third of the world's population. During the same time we leapfrogged from about 310 million dial-up, landline phones to 6 *billion* mobile phones networked by satellites and presently connecting an estimated 3.2 billion people."[39]

All of this shows that people are pretty darn clever, and we've demonstrated time and again that we can move the world when we want to. What usually stops us, though, is that first step, acknowledging that there is a problem we need to fix. It's just so, well, easy to keep on going the way we have been, to keep our heads buried in the sand, even when we know change is needed, as in, "Ah, I'll go to the gym tomorrow. Any beer left in the fridge?" That's especially the case when the detrimental changes take place gradually—not much change in that spare tire at your waistline from week to week, but one day you look in the mirror, and wow.

In the case of denying global problems, higher-level forces kick in as well. Changing the status quo can go against political or religious beliefs and can make life tougher for the small segment of the global population (often just a few individuals) that is benefiting from business as usual. For those reaping the immediate gains, there is no obvious advantage to change: they are in the "lucky" position (in the strictly selfish sense) of cashing in on the present while not only deferring the actual costs into the future but also making somebody else pay—namely, you, in a few years, and then your kids.

The forces that keep us in denial of global problems are well illustrated by the half century it has taken for the reality of global climate change to move out of academia and into mainstream society, as explained eloquently by Naomi Oreskes and Erik Conway in their 2010 book *Merchants of Doubt*.[40] In the case of climate change, special interest groups with strong political beliefs and/or financial incentives mounted a campaign to cast doubt on the concept early on, much as happened with the tobacco industry, which tried to get people to keep buying cigarettes, even though there was no question that smoking could, and often did, kill you. The financial interests in discounting climate change remain huge—from 2008 to 2012, for instance, oil giant ExxonMobil was Fortune 500's top profit-generating company for four of those five years and Chevron held down either the number two or three spot for three of those years. In 2013, ExxonMobil once again held the top spot and the Russian energy giant Gazprom was number three.[41]

The tactic of casting doubt on the whole concept of climate change and on the objectivity and motives of the scientists researching it has worked remarkably well, especially in the United States. Even by the mid-1990s, when the weight of scientific evidence was already overwhelming that burning fossil fuels, deforestation, and other human activities were changing the climate in ways people would not like, the perception in the press and the general public was that scientists disagreed about whether climate change was real at all, and assuming it was, whether people caused it.

Also contributing to the reluctance to accept the facts is the association of the climate change issue with political parties. In the United States, the idea that humans caused climate change has become linked with the Democratic Party and the growth of "Big Government." That perceived linkage in fact has been there for a long time—it was Democratic president Lyndon Johnson who told Congress, back in 1965, "This generation has altered the composition of the atmosphere on a global scale through ... a steady increase in carbon dioxide from the burning of fossil fuels."[42] When Vice President Al Gore began speaking out vociferously about climate change, the die was cast: if you had Republican values, then climate change could not be real. The U.S. political system became gridlocked—moving forward on any issue related to climate was political death. And since the United States is a chief emitter of greenhouse gases (second only to China), its past reluctance to engage in any meaningful solutions to the problem has stymied world efforts. Further exacerbating the polarization is that for some, climate change strikes at the heart of deep-seated religious beliefs: if God made the world, he made the climate, and there is nothing people can do about it one way or another.

What began to sway public opinion in the United States was the series of extreme weather events that hit in 2011 and 2012, especially Hurricane Sandy, which, as I mentioned in chapter 3, made the vision of storms flooding New York City in the movie *The Day after Tomorrow* a reality.[43] Even before that, however, surveys were showing that the majority of the American public—somewhere between 85 and 63 percent, depending on the year and the survey—recognized that climate change was happening, and at least 50 percent accepted that humans were causing it.[44] Although that was considerably less than the consensus voiced by 97 percent of the scientists who study climate change, it was apparently enough.[45] On June 25, 2013, President Barack Obama finally stated, in no uncertain terms, that climate change is real, that humans have caused it, and that, as the world's biggest emitter of greenhouse gases as well as its largest economy, the United States is going to become a leader in slowing climate change before it is too late. The comprehensive strategy he announced included cutting greenhouse emissions at their source, emplacing incentives that would lead to innovations in energy production and energy efficiency, and international cooperation. This was a huge step by a country that had stood in the way of fixing the climate problem for more than two decades. The June 25 announcement came on the heels of a joint announcement by Obama and China's President Xi Jinping that had been made a couple of weeks earlier, that both countries recognized climate change as a serious problem and were jointly initiating steps to slow it.[46]

It had been forty-eight years since Lyndon Johnson first pointed out the greenhouse gas problem to Congress, which illustrates clearly my points that altering the trajectory of dangerous global change does not happen overnight, and that the hardest step is usually getting people to accept that there is a problem. A key lesson here is that these games are played out over a human generation or two—you lose if you throw up your hands in frustration and drop out too soon. As someone once told me about effective communication, just when you are so sick of saying something that you never want to say it again is about when people start to listen. There's a lot of truth in that, as the parents of most children can attest.

Now that climate change is no longer denied by the majority of people and is recognized as a priority at the highest political levels (although in the United States, political gridlock still has to be broken), the rest becomes a matter of mechanics, at which people are very, very good. As we saw in chapter 4, the solutions are there; the optimist in me says that now that we are over the acceptance hurdle, or at least mostly over it, good change has a chance of happening fast.

Public acceptance of the climate change / energy issue also has been tough to achieve because its direct impacts on an individual person are not easy to wrap your head around unless your house is swept away in a flood or burned up in a drought-triggered wildfire. It's much easier to get traction on issues where everybody can easily see or imagine the consequences, as in starving to death. That's why, in the late 1960s, when people were just starting to argue over whether climate change was real or not, they got busy and started fixing the problem of world hunger, through what popularly became known as the Green Revolution.

The Green Revolution stands as a testament to what can be accomplished to change the course of the world when virtually everyone, from the guy on the street to people in the highest levels of government, decides the course needs to be changed. Around 1950, India's population began to grow nearly twice as fast as it had for the previous 250 years, and as a result, in the early 1960s, the country began to run out of food. India's problems reflected a similar danger that was looming for the whole world, which in large part had resulted from the rapid rise in global population that followed World War II. During and right after the war, basic sanitation was introduced to many places that had previously lacked it; as a result, death rates went way down, but birth rates did not fall correspondingly. By 1961, skyrocketing population growth coupled with back-to-back drought years and an inefficient food-growing and distribution system had brought the hunger crisis to a head in India. At the time, fields were tilled largely by hand, generally without fertilizers, and yields of essential crops like wheat were further depressed by their susceptibility to pests and diseases.

Fifteen years prior to that, an Iowa farm boy named Norman Borlaug had run into a similar situation in Mexico. Borlaug had earned a PhD in plant pathology from the University of Minnesota. From there he moved on to the DuPont chemical company, where he worked on chemical compounds deemed useful in World War II. By 1944, Borlaug had joined the Rockefeller Foundation's project to boost wheat production in Mexico, which at the time was not doing so well.

Having grown up during the Depression on a "dirt-poor" farm in Iowa, Borlaug was nevertheless shocked at what he found in Mexico. As the *New York Times* later reported, "Indeed, on first seeing the situation in Mexico for himself, Dr. Borlaug reacted with near despair. Mexican soils were depleted, the crops were ravaged by disease, yields were low and the farmers could not feed themselves, much less improve their lot by selling surplus. 'These places I've seen have clubbed my mind—they are so poor and depressing,' [Borlaug] wrote to his wife after his first extended sojourn in the country. 'I don't know what we can do to help these people, but we've got to do something.'"[47]

Help them he did, along with helping everyone else on the planet, so much so that he received a Nobel Prize for his efforts. Not bad for an Iowa farm kid who reportedly attended a one-room schoolhouse and flunked his university entrance exam.[48] Methods Borlaug pioneered in Mexico in the 1940s eventually were scaled up to accomplish what later became known as the Green Revolution, which has been credited with saving over a billion people from starvation.

The first thing Borlaug had to overcome was a reluctance to change old ways. The peasant farmers he was working with regarded metal plows as robbers of Earth's heat. Fertilizers, which Borlaug knew would help increase yields, were regarded as poison. But Borlaug wasn't deterred—in his words, "Never underestimate the little farmer's capacity to change."[49] He showed farmers how crops thrived in his experimental plots. He crossbred wheat strains to come up with varieties that were resistant to rust, which was decimating yields at the time, and to produce varieties broadly adapted to many different soil types and growing conditions.[50] Yields were increased further with the application of pesticides and by giving the wheat plants all the nitrogen they needed by applying fertilizer. A problem Borlaug ran into with fertilizer was that adding nitrogen above a certain amount caused the plants to produce very heavy seed heads—so heavy, in fact, that the stalks couldn't support them, which ruined the crop. To get around that, Borlaug and his colleagues figured out, by 1953, a way to breed wheat that had a much shorter, compact stalk that could support the enlarged seed heads.

All of this was accomplished in less than a decade and proved so successful in Mexico that the new farming approaches were transferred to the United States. As a result, the United States went from having to import half its wheat during the 1940s to becoming self-sufficient in the 1950s, and not long after that, to becoming an exporter of wheat to the world.[51]

So when the food crisis flared up in India and many other countries in the 1960s, Borlaug's techniques and technology had already been proven in North America. But there was another problem he and aid organizations had to overcome at that point—national attitudes and politics. In India and Pakistan, for example, there was great resistance to allowing the new, improved seeds and requisite fertilizer into the respective countries. Once that hurdle was cleared, the wheat harvests increased dramatically, but then, in Pakistan, things seemed to stall. As the story goes, Borlaug "was summoned back to Pakistan to explain why the Green Revolution was failing. After touring the countryside he determined yields were as high as expected. The problem was that the government had dropped its guaranteed price for wheat by 25 percent. Speculators were hoarding the crop."[52] Which led Borlaug to remark, years later, "The best plant variety is only the catalyst.... It has the potential, but you've got to know how to plant it, correct the soil infertility, and cut down competition from pests and disease. Once you've put together the jigsaw of production, you've got to further link it to economic policy that permits the little farmer to apply the technology."[53]

Changing business practices and economic incentives at the national level was the final piece of the puzzle. Before Borlaug got to India, the wheat yield there was about 800 pounds per acre; after national policies were changed to facilitate spread of the new farming methods, the yield grew to 6,000 pounds per acre. Worldwide, that success was mirrored—wheat yields went from 692 million tons produced on 1.70 billion acres of cropland in 1950 to 1.9 *billion* tons from 1.73 billion acres in 1992.[54]

The technology and methods were transferable to crops besides wheat. What made that transfer ultimately successful was formal cooperation among nations, companies, and private foundations, which resulted in the installation of facilities aimed at improving yields of various crops in several parts of the world. For instance, in 1960, coincident with Borlaug's work in India, the government of the Republic of the Philippines, the Ford Foundation, and the Rockefeller Foundation created the International Rice Research Institute (IRRI). By 1966, the IRRI had produced a new strain of rice dubbed IR8, or "Miracle Rice," which when grown under optimal conditions increased the yield per acre up to tenfold.[55]

It was that combination of recognizing a problem, changing old ways of doing things (showing farmers something that worked better), technical innovations (planting and breeding for new strains), scaling up new technology (fertilizers and pesticides), emplacing government policies (fair pricing and price guarantees) that rewarded the people producing the goods, and local through international cooperation that made the Green Revolution work so well and so fast. This is not to say that the Green Revolution answered all of our food problems forever, by any means. As noted in chapter 5, some of the techno-fixes have resulted in the pollution of land and waterways—from overuse of fertilizer and harmful pesticides, for example, and we've emphasized some crop strains so heavily that the natural genetic variation that allowed us to create those strains in the first place is on the way out. And as noted earlier, we now face a renewed round of having to change our ways in order to feed the world. Nevertheless, the lesson of the Green Revolution is clear as far as solving world problems goes: it's doable, and moreover, it's doable within three to five decades.

The same applies to our economic woes. In fact, some two hundred major corporations—each of whose net worth exceeds the annual budget of many countries—have joined forces as the World Business Council for Sustainable Development, with the express purpose of coming up with business solutions that do not have harmful environmental impacts.[56] Among the participating businesses are such giants as Toyota, Boeing, Alcoa, Syngenta, Accenture, Volkswagen, Duke Energy, Infosys, PricewaterhouseCoopers, GDF Suez, and Weyerhaeuser.

The CEOs of twenty-nine of those companies, in dialogue with the rest, spent eighteen months coming up with a plan they call Vision 2050. These CEOs, remember, are the people whose bottom lines depend on developing workable business models with definable milestones that build toward big payoffs. Their goal in producing Vision 2050 was to articulate, with suitable milestones, a "new agenda for business laying out a pathway to a world in which nine billion people can live well, and within the planet's resources, by mid-century." This is not pie in the sky for these CEOs. It is where they want their companies and the world to be in thirty-five years, and they've embarked on a path toward it with a plan called Action 2020.[57]

Vision 2050's cornerstones include most of the things I've mentioned earlier that will be crucial to staving off the Sixth Mass Extinction: incorporating the costs of ecosystem services into the marketplace; doubling agricultural output without increasing the amount of land or water used; halving deforestation and increasing yields from planted forests; shifting to low-carbon energy systems; and improving energy efficiency. Targets for turning around the biodiversity decline are specified for each of the next few decades. For the 2010s, for instance, the targets include achieving widespread literacy about the value of biodiversity; increasing investments in rural communities in and around protected regions; business, nongovernmental organizations, and governments agreeing on the valuation of various ecosystem services; and seed banks and botanical gardens all over the world working together to ensure preservation of plant genetic diversity. The goals for the 2020s are to have fair pricing in place for ecosystem services; enlarge and enhance critical habitats; and work to ensure that at least some ecosystem services are the most lucrative investments on the market. Of course, not all corporations share these values—ExxonMobil is a notable holdout, for example—but the direction in which the companies in the council are going, especially as laid out in Vision 2050 and Action 2020, shows that the barons of industry are ready to change the world for the better if consumers give them a signal that that is the direction we want to go in as well.[58]

So here we stand. Much of the general public and the highest levels of government, including those in the United States and China, recognize the necessity of reducing the use of fossil fuels and slowing climate change. Likewise, we know food security is a looming problem and we know what to do about it. And as substantiated by the Vision 2050 and Action 2020 plans, big business is getting on board with actions needed to stop biodiversity loss.

That means that where we stand now with respect to solving the Sixth Mass Extinction is very much where Norman Borlaug stood at the beginning of the Green Revolution. Back then, I'd venture to say, he would have been hard-pressed to find anyone who was willing to watch a billion people starve to death. By the same token, I think that today you wouldn't find many people who would be happy to watch three-quarters of the worlds' mammals, birds, reptiles, and amphibians die out. And just as Borlaug had new ideas and access to proven technology that was ready to be implemented, so too do we have new innovations ready to go in the arenas of energy production, food security, and economic viability. Finally, if individuals, local governments, and nations were able to change old ways of doing business and cooperate to accomplish the Green Revolution, there's no reason they can't do the same to avoid the Sixth Mass Extinction.

Naturally, it won't be easy—changing course never is. What it will take, more than anything else, is the conscious desire of all of us to make a better world translated into meaningful action. Taking action is where it's all too easy to stall out, especially if you start thinking that you have to change the world all by yourself. You don't—in fact, it's multiplying the things each one of us does by billions of people that got us into the mess, and that's exactly what will get us out. The other thing to remember is that not everybody can, or even should, take the same actions. The menu of potential actions for a poor laborer in an undeveloped country is very different from that for the venture capitalist or the average blue-collar worker in an urban area; therefore, what you can and should do is going to very much depend on your circumstances.

I'm guessing that most of you reading this book consider yourselves to be average people just trying to get by in a place like the United States or Europe, where the standard of living is pretty high. For you, the first thing to do is easy: spread the message among your friends, neighbors, and coworkers that the extinction crisis is real, dire, and fixable, and that the fix will make the world a much better place for ourselves and our kids, a place with fewer floods and heat waves, less famine, and more stable economies. The second thing takes a little more effort, but not much more: pay attention to what you eat. Avoid foods with palm oil—remember, Indonesian and other rainforests are being cut down to make room to grow palm oil plants. When possible, buy local foods (reduced transportation distance reduces greenhouse gas contribution, among other things), and plan meals so that you don't throw away food. Use a guide like the Monterey Bay Aquarium's Seafood Watch list (see chapter 5) to make sure the fish you buy are sustainable.[59] Eat a little less meat, and when you do eat meat, go for the grass-fed rather than the feedlot-fattened variety. The third way you can contribute is to lower your carbon footprint, even a little. Take advantage of the free audits that many utility companies provide to identify inefficiencies in home energy use that can lead to easy savings; trip-chain (go from one errand to the next) rather than taking several short round trips when you use your car for errands; recycle; and when you can, walk or bike instead of driving.

These are, of course, things you have heard over and over and may be sick of hearing; if you are already doing them, great. But you might be surprised at how many people are not; if you're not, it's time to start. All of these little actions, when multiplied by hundreds of millions of the world's most prolific consumers, add up to huge results. It doesn't matter if you are a mechanic, a grocery store clerk, a student, a teacher, a banker, a retiree, or anything else—these are things everybody can do to make a difference.

It's also crucial that you put into public office leaders who recognize the importance of switching from a fossil-fuel energy system to a carbon-neutral one, who see the necessity of growing crops more efficiently, and who are willing to provide economic incentives and legal teeth to protect threatened species. Take the time to find out candidates' stances on these issues before you vote. In addition, you have the power to tell the CEOs who are guiding the future of business that you won't tolerate robbing nature's bank, by choosing to spend your money on environmentally friendly companies.[60] By this point I hope it goes without saying, but never, ever buy items made from ivory, rhino horn, or any other parts of those species and the many others that are under threat of extinction. When in doubt, pull out your smart phone and check the list of species protected under the Convention on International Trade in Endangered Species of Wild Fauna and Flora.[61]

Some of you are in a position to do more than others. If you happen to be CEO of a company, you have more responsibility, as well as more power, to prevent extinctions. Your chief contribution in that respect is to develop business models that will both better people's lives today and ensure that their kids' and grandkids' lives won't be worse. If your supply chain contributes to habitat destruction or otherwise impacts endangered species, change the supply chain or change the product. Ultimately, you need to design and provide products that end up, as they go from cradle to grave, with a net environmental footprint of zero as far as critical ecosystem services and survival of species go. You probably also need a tax break. Get it by helping organizations that are keeping threatened species alive.

If your influence is in government, your role is to educate your constituencies and work with others elected to lead—both within and across your political affiliations and jurisdictions—to emplace the incentives and policies essential to promoting carbon-neutral energy and more efficient and environmentally beneficial agriculture, and that value ecosystem services and respect both human life and that of other species. There are political realities, for sure—like the dysfunction that grips Washington, DC, as I write this—but there are always doors of opportunity as well, as evidenced by recent accords that bypass national political gridlock and effectively push global policies in productive directions. Notable examples include a September 2013 agreement between the world's eighth- and second-largest economies, California and China, respectively, to significantly reduce greenhouse gas emissions and develop green technologies.[62] Just about six weeks after the California-China memorandum of understanding was signed, the elected leaders of California, Oregon, Washington, and British Columbia signed an agreement to battle climate change with concrete actions like facilitating solar and wind projects and better integrating them into the electric power grid, informing policy with science, and cooperating "with national and sub-national governments around the world to press for an international agreement on climate change in 2015."[63] A few days later, President Obama appointed some of those same signatories (the governors of the three U.S. states) plus five other governors to the Task Force on Climate Preparedness and Resilience, which is charged with providing advice on how the federal government "can respond to the needs of communities nationwide that are dealing with the impacts of climate change."[64] These actions exemplify two things. First, similarly bold steps can be taken by any number of sub-national and national leaders, not only to deal with climate change, but also to guide the future of food and money as they relate to the extinction crisis and human well-being. Second, the politicians who are taking these decisive actions show us how leaders can actually, well, lead.

Some of you, including those who are presently students in universities and high schools, are not world leaders yet, but you will be in just a few years. Your action items are to a large extent different from, and in some ways even more important than, those that apply to your parents' generation. You're at the beginning and looking to the future. You see the world as it is today, not as it was twenty or thirty or fifty years ago, and that means you're not focusing on how things used to be and how hard it is to change business as usual. Instead, those of you with whom I've had the privilege of interacting are ready to make things how you want them to be; you see the opportunities and are ready to grab them. Here's how. First, make a career out of something you enjoy—that's what will make you good at it. When you decide what that is, practice it in a way that takes into account that no longer is there any "local"; you depend on goods and services from, and successful interactions with, people who live on the other side of the world, and they depend on you. While you move through your career, use that underlying reality to your advantage—there are adventures to be had and money to be made while at the same time helping to solve the world's power, food, and money problems and the resulting extinction crisis.

The specifics will, of course, differ depending on which career we're talking about. If you are an architect, the wave of the future is in designing buildings and communities that are not only attractive and functional but also energy efficient, that have a minimal footprint as far as materials and maintenance are concerned, and that integrate nicely into natural habitats. If you are an engineer, you could make billions by improving carbon-neutral energy systems. If you are a teacher, it's all about instilling in your students the knowledge that their lives are connected to the lives of people who speak a language they probably don't even know, and teaching them not just facts but how to think for themselves and solve problems. If you go into law, it's going to be about things like forging international agreements, coming up with ways to price ecosystem services, and dealing with the complexities of environmental protection. If it's investment banking, it's about valuing and investing in natural capital and in green technologies that help developing countries leapfrog over outdated approaches that have proven to be environmentally destructive. And if your forte is something like languages, philosophy, literature, or art, your role is going to be particularly important. People who can translate the nuances that keep different cultures from understanding each other, and those who can come up with ways to communicate complexities so that everyone can understand them, will be essential in building a well-functioning global ecosystem.

A special word to budding scientists, since that's what I've been for most of my life: if science is your thing, make sure you're researching a problem that's worth spending time on. When you make that big breakthrough, make sure you communicate it outside the ivory tower; no longer is it enough to bury your discoveries in an academic journal. Don't give up until people hear you—it could, and probably will, take years. Persistence pays off: for decades scientists have been discussing the issues I've covered in this book at professional conferences and publishing about them in venues read mainly by other scientists, but only very recently has the need for active solutions begun to drift onto the public's radar screen as scientists become better at communicating their important discoveries to non-scientists.[65] Ironically, many scientists I know—even biologists—still think that global issues are somebody else's problem to solve and that their job stops at reporting an arcane finding in an obscure scientific journal. Don't fall into that trap. The list of what you can do profession by profession goes on, but you get the idea: make the choices in your job that contribute to the solutions, instead of the choices that cause more problems.

But what about people in developing countries, places like India, Pakistan, and rural China, who are rightly striving to gain the modern conveniences that I, and most of you reading this book, take for granted? The people working their way up the economic ladder in developing countries account for more than a third of the world's population. How can they become part of the solution instead of repeating the mistakes we've made? If you are one of those people, building schools and making sure girls as well as boys attend is a good place to start. Access for all to education, wherever it is embraced, has time and again resulted in more global connection, more economic opportunities, and reduced family size, all of which are necessary for decreasing the global human footprint, which in turn is essential for keeping other species on Earth.[66] The second concrete action for those in developing countries is to use the international community relentlessly—both its funds and its expertise—to help craft and emplace local solutions that bypass existing ecologically destructive technologies. Examples of "leap-frog" solutions include the proliferation of cell phones that make landlines obsolete and solar lights that don't require an expensive energy grid.[67] Finally, if you're one of those people in developing nations, don't be lulled by promises of help and favors now that in the long run make you subject to outside whims you have no control over. Make it an equal partnership: by your sheer numbers, you wield tremendous might in the global context.

A critical responsibility falls on the shoulders of just a few people in developing countries: accepting that they are the front-line guardians of many of the iconic species that are about to disappear. If you are lucky enough to live among threatened species that many of us only read about, your job, if you want to make a difference, is to protect them. Those species really don't live anywhere else, even though you've grown up with them, and if you don't keep them alive, no one else can. They, and the habitats they live in, are very likely your chief assets now and can be the ticket to a better future for you, your children, and your grandchildren. Don't let anyone steal those assets from you, and don't be complicit in wasting them yourself, because if you do, you'll suddenly find that they're gone forever.

That's a long list of things we all can do, and it's by no means exhaustive; we each have a role to play. It's daunting but it's feasible, and, if we all get busy, we can turn the tide just in time. Taking collective action will almost certainly back us away from the brink of the Sixth Mass Extinction. In that sense, we're poised at another kind of brink, the brink of opportunity. Which precipice we go over, and which one we decide to pull back from, will be a game changer, one way or another. It will define not only our generation's place in history but also the future of humankind.

Acknowledgments

I wrote this book while on sabbatical at Stanford University as Cox Visiting Professor in the Department of Environmental Earth System Science. I thank Rob Dunbar for nominating me for that fellowship and Pamela Matson for her part in providing the requisite funding and facilities. I also am grateful to my home institution, the University of California at Berkeley, and the departments with which I am affiliated there—the Department of Integrative Biology, the Museum of Paleontology, and the Museum of Vertebrate Zoology—for granting the sabbatical I needed to delve deeply into this subject. Part of the work was done in connection with one of my research projects on extinctions that was funded by the National Science Foundation Sedimentary Geology and Paleobiology Program, whose support I gratefully acknowledge. I appreciate Blake Edgar's willingness to take on the book at University of California Press, and his guidance. I also thank Merrik Bush-Pirkle, production editor Dore Brown and copy editor Jan Spauschus, and the rest of the able UC Press staff for their part in the editorial and publication process. Jonathan Cobb provided valuable advice and ideas in the conceptual stages of the book project.

Thanks are due to several people who were kind enough to read various chapters or parts thereof and give me constructive comments: Walter Alvarez, Bill Clemens, Harry Greene, Robert Horn, Kaitlin Maguire, Michael McKinney, Roz Naylor, Jon Payne, Stuart Pimm, Sujith Ravi, Allison Stegner, Mark Tercek, and Susumu Tomiya. Sally Benson pointed me toward helpful literature on energy production and consumption. I take full credit for any mistakes I may have made.

Above all, I owe much to my wife, Liz Hadly, not only for putting up with me while I was writing under deadline, but especially for continually inspiring me through our discussions and her ideas, and by her own example of trying to make the world a better place. Our daughters, Emma and Clara Barnosky, continue to keep me looking toward the future, hopeful that it will be a good one.

Notes

1. THE LAST ONES STANDING

1. Charles Haskins Townsend, 1925, "The Galapagos Tortoises in Their Relation to the Whaling Industry," New York Aquarium Nature Series, reprinted from *Zoologica* 4(3):55–135, http://mysite.du.edu/~ttyler/ploughboy/townsendgaltort.htm as of September 4, 2012.

2. Ibid.

3. BBC World News, June 24, 2012, "Last Pinta Giant Tortoise Lonesome George Dies," www.bbc.co.uk/news/world-18574279 as of September 4, 2012.

4. Charles Darwin, 1839, *The Voyage of the Beagle, Complete and Unabridged Edition with a Special Introduction by Walter Sullivan, Printed Especially for Members of the Oceanic Society* (New York: Bantam; this edition published in 1972), p. 323; also available at www.literature.org/authors/darwin-charles/the-voyage-of-the-beagle/.

5. Ibid., p. 341.

6. BBC News, September 12, 2012, www.bbc.co.uk/news/uk-scotland-highlands-islands-19569538, last accessed September 14, 2012.

7. For further explanation and background, see International Union for Conservation of Nature, "IUCN Red List Categories and Criteria," www.iucnredlist.org/technical-documents/categories-and-criteria/2001-categories-criteria, last accessed September 19, 2012; and Georgina M. Mace, Nigel J. Collar, Kevin J. Gaston, Craig Hilton-Taylor, H. Resit Akçakaya, Nigel Leader-Williams, E.J. Milner-Gulland, and Simon N. Stuart, 2008, "Quantification of Extinction Risk: IUCN's System for Classifying Threatened Species," *Conservation Biology* 22:1424–1442.

8. "Tanzania: Serengeti/Mara East African Synergy or Primitive Rivalry?" *AllAfrica Tanzania Daily News,* August 18, 2012, http://allafrica.com/stories/201208180382.html, last accessed September 15, 2012.

9. "Cameroon Losing Battle to Save Elephants from Poachers," *USA Today,* March 16, 2012, www.usatoday.com/news/world/environment/story/2012-03-16/cameroon-elephants-poaching/53564500/1, last accessed September 15, 2012.

10. Jeffrey Gettleman, "Elephants Dying in Epic Frenzy as Ivory Fuels Wars and Profits," *New York Times,* September 3, 2012, www.nytimes.com/2012/09/04/world/africa/africas-elephants-are-being-slaughtered-in-poaching-frenzy.html, last accessed September 15, 2012.

11. "Hwange Cyanide Elephant Death Toll Reaches 81," *New Zimbabwe,* September 24, 2013, http://allafrica.com/stories/201309250121.html, last accessed October 14, 2013.

12. As of October 14, 2013, www.iucnredlist.org/.

13. For the breakdown of threatened species per taxonomic group, go to www.iucnredlist.org/; on the top menu, click on the threat category you are interested in, and then click on "Taxonomy" on the left sidebar to choose the taxonomic group of interest. The number of species in that category and taxonomic group will appear in parentheses next to the name of the group.

14. Page 1428 in Georgina M. Mace, Nigel J. Collar, Kevin J. Gaston, Craig Hilton-Taylor, H. Resit Akçakaya, Nigel Leader-Williams, E.J. Milner-Gulland, and Simon N. Stuart, 2008, "Quantification of Extinction Risk: IUCN's System for Classifying Threatened Species," *Conservation Biology* 22:1424–1442.

15. The Convention on International Trade in Endangered Species of Wild Fauna and Flora (CITES) is an international agreement to mitigate international commerce in specimens of wild plants and animals.. Participating countries adhere to the convention voluntarily and basically agree to prevent the transport of listed species or their products across their boundaries. The convention began as a resolution of IUCN members in 1963. Its text was agreed upon by member countries in 1973 and it entered into force in 1975. See www.cites.org/eng/disc/what.php for additional information (website last accessed September 19, 2012).

2. IT'S NOT TOO LATE (YET)

1. The James E. Martin Paleontology Research Laboratory at South Dakota School of Mines was named in Jim's honor upon his retirement. The crew also included Nina Jablonski, who later became a leading primatologist,

paleontologist, and author, first at the California Academy of Sciences and now at Penn State University. My good friend Steve Nelson, paleontologist extraordinaire when he wasn't saving people or buildings during his day job as a Seattle fireman, rounded out our team, along with his two daughters, his mom, and another volunteer, Bob Talbot.

2. To be precise, the Pierre Shale Group contains multiple gray shale rock formations; around Edgemont, the mosasaurs we searched for were in the Sharon Springs Formation, as defined by James E. Martin, Janet L. Bertog, and David C. Parris, 2007, "Revised Lithostratigraphy of the Lower Pierre Shale Group (Campanian) of Central South Dakota, Including Newly Designated Members," *Geological Society of America Special Papers* 427:9–21, doi:10.1130/2007.2427(02).

3. To geologists, this is known as the Law of Superposition. It holds because sedimentary rocks—those that are made when tiny particles of sand or silt or clay erode off of preexisting rocks or precipitate out of water—build up in layers. The first layer has to be there before the second layer can accumulate on top of it, and so on. This regular order is preserved in the rock record unless some truly remarkable tectonic activity flips the whole sequence.

4. "Absolute" ages of rocks, and therefore of the fossils the rocks encase, are determined by looking at the ratios of various isotopes that decay, parent isotope into daughter isotope, at known rates. These isotopes are found in elements such as potassium, argon, uranium, and carbon. The isotopes start to decay as soon as the mineral that contains them solidifies out of magma. By knowing the rate of decay and measuring how much of the daughter isotope has accumulated relative to the parent isotope, it is possible to calculate how many years ago the mineral formed, within certain margins of error. The margins for older rocks tend to be plus or minus hundreds of thousands of years (not bad when you're talking about hundreds of millions of years). For younger deposits dated by the radiocarbon method, the error bars can be as small as a few tens of years.

5. Older literature calls it the K-T, with T referring to a period called the Tertiary, out of which the Paleogene was carved.

6. Luis W. Alvarez, Walter Alvarez, Frank Asaro, and Helen V. Michel, 1980, "Extraterrestrial Cause for the Cretaceous-Tertiary Extinction: Experimental Results and Theoretical Interpretation," *Science* 208:1095–1108.

7. Much later, the Law of Uniformitarianism was interpreted more broadly, recognizing the difference between uniformity of rate and uniformity of process. In the latter context, "process" has been taken to mean pretty much anything that has happened in the creation and evolution of the universe,

so in that case an asteroid impact would certainly not be inconsistent with the law.

8. J. David Archibald and William A. Clemens, 1982, "Late Cretaceous Extinctions," *American Scientist* 70:377–385.

9. Peter M. Sheehan, David E. Fastovsky, Raymond G. Hoffman, Claudia B. Berghaus, and Diane L. Gabriel, 1991, "Sudden Extinction of the Dinosaurs: Latest Cretaceous, Upper Great Plains, U.S.A.," *Science* 254:835–839.

10. Alan R. Hildebrand, Glen T. Penfield, David A. Kring, Mark Pilkington, Antonio Camargo Z., Stein B. Jacobsen, and William V. Boynton, 1992, "Chicxulub Crater: A Possible Cretaceous/Tertiary Boundary Impact Crater on the Yucátan Peninsula, Mexico," *Geology* 19:867–871. Alan Hildebrand was a PhD student studying planetary sciences at the University of Arizona when he did this work; later he joined the Canada Geological Survey and the faculty of the University of Calgary.

11. Douglas S. Robertson, Malcolm C. McKenna, Owen B. Toon, Sylvia Hope, and Jason A. Lillegraven, 2004, "Survival in the First Hours of the Cenozoic," *Geological Society of America Bulletin* 116:760–768.

12. Peter Schulte, Laia Alegret, Ignacio Arenillas, José A. Arz, Penny J. Barton, Paul R. Bown, et al., 2010, "The Chicxulub Asteroid Impact and Mass Extinction at the Cretaceous-Paleogene Boundary," *Science* 327:1214–1218.

13. J. David Archibald, W. A. Clemens, Kevin Padian, Timothy Rowe, Norman Macleod, Paul M. Barrett, et al., 2010, "Cretaceous Extinctions: Multiple Causes," *Science* 328:973.

14. Paul R. Renne, Alan L. Deino, Frederik J. Hilgen, Klaudia F. Kuiper, Darren F. Mark, William S. Mitchell III, Leah E. Morgan, Roland Mundil, and Jan Smit, 2013, "Time Scales of Critical Events around the Cretaceous-Paleogene Boundary," *Science* 339:684–687.

15. That 1,000-plus includes species we drove to extinction over the past 50,000 years, as well as the approximately 800 we have killed off in the last 500 years.

16. Anthony D. Barnosky, Nicholas Matzke, Susumu Tomiya, Guin Wogan, Brian Swartz, Tiago Quental, et al., 2011, "Has the Earth's Sixth Mass Extinction Already Arrived?" *Nature* 471:51–57.

17. The percentage for mammals follows from a count of 429 mammal species extinctions in the last 50,000 years as compiled from paleontological literature and IUCN counts. David Steadman estimated some 2,000 species of birds disappeared from Pacific islands when people arrived, which, when added to the global count of species that went extinct over the past five

centuries and presently extant species, works out to about a 17 percent species loss. See David W. Steadman, 1995, "Prehistoric Extinctions of Pacific Island Birds: Biodiversity Meets Zooarchaeology," *Science* 267:1123–1131.

18. Simon N. Stuart, Janice S. Chanson, Neil A. Cox, Bruce E. Young, Ana S.L. Rodrigues, Debra L. Fischman, and Robert W. Waller, 2004, "Status and Trends of Amphibian Declines and Extinctions Worldwide," *Science* 306:1783–1786.

3. A PERFECT STORM

1. The original "perfect storm" was the violent tempest that hit New England on November 1, 1991, when Hurricane Grace was absorbed by a nor'easter at the edge of a cold front. The 1997 book and the 2000 movie *The Perfect Storm* were based on this event and launched the phrase into the popular vernacular.

2. To carry the metaphor further, your normal body temperature is around 98.6 degrees Fahrenheit. If you are running a fever of only 0.4 degree—a body temperature of 99.0—you feel it. Two degrees higher, near 101, you feel very ill. Six degrees higher, near 105, and you die if it doesn't come down quickly. The point is that even a few degrees above normal make a tremendous difference in the bigger scheme of things.

3. Some studies suggest the extinction took as long as two million years to unfold, but recent work brackets it within 60,000 years. The time span may have been much shorter, but due to the vagaries of geological dating, 60,000 years is the best that can be done in terms of figuring out how long the extinction actually took. See J.L. Payne and M.E. Clapham, 2012, "End-Permian Mass Extinction in the Oceans: An Ancient Analog for the Twenty-First Century?" *Annual Review of Earth and Planetary Sciences* 40:89–111; S. D. Burgess, S. Bowring, and S.-Z. Shen, 2014, "High-Precision Timeline for Earth's Most Severe Extinction," *Proceedings of the National Academy of Sciences* 111: 3316–3321.

4. For more details, see ibid.

5. These were the early diapsid animals, which are united evolutionarily by having two openings for muscle attachments toward the back of the skull. The name "diapsid" means "two-arch," in reference to two horizontal arches at the rear of the skull, each of which forms a lower bony border for one of the muscle-attachment openings.

6. Douglas H. Erwin, 1994, "The Permo-Triassic Extinction," *Nature* 367:231–236.

7. Zhong-Qiang Chen and Michael J. Benton, 2012, "The Timing and Pattern of Biotic Recovery following the End-Permian Mass Extinction," *Nature Geoscience* 5:375–383, doi:10.1038/NGEO1475.

8. The notation $\delta^{13}C_{carb}$ actually expresses the amount that the ratio of $^{13}C/^{12}C$ in carbonate sediments changed in relation to the ratio of $^{13}C/^{12}C$ in an agreed-upon standard specimen, which is the ratio in a certain Cretaceous fossil called the Pee Dee Belemnite. The subscript "carb" specifies the analysis was done on carbonate. One can also determine the ratios of carbon isotopes from materials other than carbonate, for instance, from organic carbon, in which case $\delta^{13}C_{org}$ would be the notation.

9. Mordeckai Magaritz, Richard Bärr, Aymon Baud, and William T. Holser, 1988, "The Carbon-Isotope Shift at the Permian/Triassic Boundary in the Southern Alps Is Gradual," *Nature* 331:337–339.

10. William T. Holser, Hans-Peter Schönlaub, Moses Attrep, Jr., Klaus Boeckelmann, Peter Klein, Mordeckai Magaritz, et al., 1989, "A Unique Geochemical Record at the Permian/Triassic Boundary," *Nature* 337:39–44.

11. Jonathan L. Payne, Daniel J. Lehrmann, Jiayong Wei, Michael J. Orchard, Daniel P. Schrag, and Andrew H. Knoll, 2004, "Large Perturbations of the Carbon Cycle during Recovery from the End-Permian Extinction," *Science* 305:506–509.

12. M.R. Rampino and K. Caldeira, 2005, "Major Perturbation of Ocean Chemistry and a 'Strangelove Ocean' after the End-Permian Mass Extinction," *Terra Nova* 17:554–559.

13. J.L. Payne and M.E. Clapham, 2012, "End-Permian Mass Extinction in the Oceans: An Ancient Analog for the Twenty-First Century?" *Annual Review of Earth and Planetary Sciences* 40:89–111.

14. An interesting aside: the word "traps" in this case derives from the Swedish word "trappa" or "trap," which means "stairs" and describes the stair-step appearance the lava flows give to the landscape.

15. The lower estimate (11,000 GtC) comes from Uwe Brand, Renato Posenato, Rosemarie Came, Hagit Affek, Lucia Angiolini, Karem Azmy, and Enzo Farabegoli, 2012, "The End-Permian Mass Extinction: A Rapid Volcanic CO_2 and CH_4-Climatic Catastrophe," *Chemical Geology* 322/323:121–144. The upper estimate (30,000 GtC) comes from J.L. Payne and M.E. Clapham, 2012, "End-Permian Mass Extinction in the Oceans: An Ancient Analog for the Twenty-First Century?" *Annual Review of Earth and Planetary Sciences* 40:89–111. Additional plausible sources of CO_2 are the burning or decay of organic matter and the release of biogenic methane. The former would include burning of coal

seams and vegetation set on fire by volcanic eruptions, as well as the decomposition of vegetation, animals, and phytoplankton that were killed as the extinction event went into full swing. The latter comes from frozen bubbles of methane in ocean sediments that can suddenly melt if the ocean bottom warms enough.

16. The Intergovernmental Panel on Climate Change, or IPCC, models the range of global temperature rise that would result under different assumptions of greenhouse gas emissions, assumptions that in turn are based on alternative possibilities of economic growth, the speed at which fossil fuels might be replaced by carbon-neutral energy, and so on. There are two IPCC reports that are commonly cited, the 2007 report and the 2013/2014 report. The 2007 IPCC report is abbreviated AR4 (IPCC, 2007, "Summary for Policymakers," in *Climate Change 2007: The Physical Science Basis,* ed. S. Solomon, D. Qin, M. Manning, Z. Chen, M. Marquis, K.B. Averyt, M. Tignor, and H.L. Miller [Cambridge: Cambridge University Press], online at www.ipcc.ch/pdf/assessment-report/ar4/wg1/ar4-wg1-spm.pdf). The 2013/2014 report is abbreviated AR5 (IPCC, 2013, "Summary for Policymakers," in *Climate Change 2013: The Physical Science Basis,* ed. T.F. Stocker, D. Qin, G.-K. Plattner, M. Tignor, S.K. Allen, J. Boschung, A. Nauels, Y. Xia, V. Bex, and P.M. Midgley [Cambridge: Cambridge University Press], online at www.ipcc.ch/report/ar5/wg1/docs/WGIAR5_SPM_brochure_en.pdf). The two reports use slightly different ways to discuss future emissions trajectories. In the AR4 report, the alternative future scenarios are abbreviated with various letter-number combinations, such as B1, B2, A2, and A1F1. I've listed those four examples in order of increasing emissions. The A2 scenario indicates at least a 66 percent chance of a 5.4-degree Celsius increase. However, if we continue emitting greenhouse gases at the percapita rate that has prevailed for the past few years, we will follow a warming trajectory even steeper than the A2 trajectory, according to Steven J. Davis, Long Ca, Ken Caldeira, and Martin I. Hoffert, 2013, "Rethinking Wedges," *Environmental Research Letters* 8(1):011001, doi:10.1088/1748-9326/8/1/011001. The scenario that somewhat parallels our present trajectory is the A1F1 scenario, which indicates at least a 66 percent probability that the mean global temperature will rise 6.4 degrees Celsius by 2100. The more recently published AR5 report (2013/2014) calls the potentially different emissions scenarios Representative Concentration Pathways, abbreviated RCP; the RCP8.5 projection suggests at least a 66 percent probability of a 4.8-degree Celsius rise.

17. As of January 2013 there were about 395 parts per million (ppm) of CO_2 in our atmosphere, which equates to about 841 gigatons of carbon (abbreviated

GtC). That value is up about 115 ppm (or 245 Gtc) from where it was before humans starting adding lots of CO_2 by burning fossil fuels. This is much below the estimates for the Permian atmosphere, which range from 1,810 to 50,000 GtC. Part of the explanation of how the Permian world could have had such high CO_2 concentrations yet still be close in average temperature to today's world is that the sun was somewhat weaker in the late Permian, when it was 252 million years younger. As stars like the sun age, they become brighter and hotter until they run out of hydrogen fuel and collapse into white dwarfs. Luckily for us, the sun still has about five billion years before this happens.

18. For the period 1901 to 2000, Earth's average surface temperature was about 57 degrees Fahrenheit (13.9 degrees Celsius).

19. This quotation and the following two are from William G. Melton, Jr., 1972, "The Bear Gulch Limestone and the First Conodont Bearing Animals," in *Field Guide for the 21st Annual Field Conference of the Montana Geological Society* (Montana Geological Society), pp. 65–68.

20. For an earlier study, see M.R. Rampino and K. Caldeira, 2005, "Major Perturbation of Ocean Chemistry and a 'Strangelove Ocean' after the End-Permian Mass Extinction," *Terra Nova* 17:554–559.

21. Yadong Sun, Michael M. Joachimski, Paul B. Wignall, Chunbo Yan, Yanlong Chen, Haishui Jiang, Lina Wang, Xulong Lai, 2012, "Lethally Hot Temperatures during the Early Triassic Greenhouse," *Science* 338:366–370.

22. J.L. Payne and M.E. Clapham, 2012, "End-Permian Mass Extinction in the Oceans: An Ancient Analog for the Twenty-First Century?" *Annual Review of Earth and Planetary Sciences* 40:89–111.

23. S. Solomon, D. Battisti, S. Doney, K. Hayhoe, I.M. Held, D.P. Lettenmaier, et al., 2011, *Climate Stabilization Targets: Emissions, Concentrations, and Impacts of Decades to Millennia* (Washington, DC: National Academies Press).

24. The future emissions scenarios of the IPCC suggest that the mid-range scenarios would put between 1,100 and 1,800 GtC into the atmosphere by 2100, which translates to an increase of between 516 and 845 ppm. See figure 4 in IPCC, 2000, "Emissions Scenarios: Summary for Policymakers," www.ipcc.ch/pdf/special-reports/spm/sres-en.pdf. The IPCC's AR5 reports similar future emissions; see figure SPM.10 in IPCC, 2013, "Summary for Policymakers," in *Climate Change 2013: The Physical Science Basis,* ed. T.F. Stocker, D. Qin, G.-K. Plattner, M. Tignor, S.K. Allen, J. Boschung, A. Nauels, Y. Xia, V. Bex, and P.M. Midgley (Cambridge: Cambridge University Press), online at www.ipcc.ch/report/ar5/wg1/docs/WGIAR5_SPM_brochure_en.pdf.

25. B. Hönisch, A. Ridgwell, D.N. Schmidt, E. Thomas, S.J. Gibbs, A. Sluijs, et al., 2012, "The Geological Record of Ocean Acidification," *Science* 335:1058–1063.

26. A. Barton, B. Hales, G.G. Waldbusser, C. Langdon, and R.A. Feely, 2012, "The Pacific Oyster, *Crassostrea gigas,* Shows Negative Correlation to Naturally Elevated Carbon Dioxide Levels: Implications for Near-Term Ocean Acidification Effects," *Limnology and Oceanography* 57(3):698–710.

27. H. Baumann, S.C. Talmage, and C.J. Gobler, 2011, "Reduced Early Life Growth and Survival in a Fish in Direct Response to Increased Carbon Dioxide," *Nature Climate Change* 2(1):38–41; A.Y. Frommel, R. Maneja, D. Lowe, A.M. Malzahn, A.J. Geffen, A. Folkvord, U. Piatkowski, T.B.H. Reusch, and C. Clemmesen, 2011, "Severe Tissue Damage in Atlantic Cod Larvae under Increasing Ocean Acidification," *Nature Climate Change* 2(1):42–46; M.C.O. Ferrari, R.P. Manassa, D.L. Dixson, P.L. Munday, M.I. McCormick, M.G. Meekan, A. Sih, and D.P. Chivers, 2012, "Effects of Ocean Acidification on Learning in Coral Reef Fishes," *PLoS ONE* 7(2):e31478, doi:31410.31371/journal.pone.0031478.

28. J.L. Payne and M.E. Clapham, 2012, "End-Permian Mass Extinction in the Oceans: An Ancient Analog for the Twenty-First Century?" *Annual Review of Earth and Planetary Sciences* 40:89–111.

29. R.J. Diaz and R. Rosenberg, 2008, "Spreading Dead Zones and Consequences for Marine Ecosystems," *Science* 321:926–929.

30. P.G. Harnik, H.K. Lotze, S.C. Anderson, Z.V. Finkel, S. Finnegan, D.R. Lindberg, et al., 2012, "Extinctions in Ancient and Modern Seas," *Trends in Ecology & Evolution* 27(11):608–617.

4. POWER

1. A joule is a derived unit that expresses amount of energy, work, or heat. In terms of energy, it is the work required to produce one watt of power for one second. An exajoule (EJ) is 10^{18} joules. For perspective, the Tōhoku earthquake that occurred in Japan in 2011 and that triggered the devastating tsunami and destruction of nuclear facilities released 1.41 EJ of energy.

2. The value for total solar energy per year is from Vaclav Smil, 2006, *Energy: A Beginner's Guide* (Oxford: Oneworld Publishers). The value of 550 EJ produced and consumed by humans in 2010 is from The Oil Drum, www.theoildrum.com/node/8936.

3. For the 46 Pg estimate, see Stephen Del Grosso, William Parton, Thomas Stohlgren, Daolan Zheng, Dominique Bachelet, Stephen Prince, Kathy Hibbard,

and Richard Olson, 2008, "Global Potential Net Primary Production Predicted from Vegetation Class, Precipitation, and Temperature," *Ecology* 89:2117–2126. The 66.5 Pg estimate is from Helmut Haberl, K. Heinz Erb, Fridolin Krausmann, Veronika Gaube, Alberte Bondeau, Christoph Plutzar, Simone Gingrich, Wolfgang Lucht, and Marina Fischer-Kowalski, 2007, "Quantifying and Mapping the Human Appropriation of Net Primary Production in Earth's Terrestrial Ecosystems," *Proceedings of the National Academy of Sciences* 104:12942–12947. For 53.1 Pg, see W. Kolby Smith, Maosheng Zhao, and Steven W. Running, 2012, "Global Bioenergy Capacity as Constrained by Observed Biospheric Productivity Rates," *Bioscience* 62:911–922. I use this estimate (53.1 Pg) because it's a midrange estimate and relies on more recently acquired data than the other two.

4. See W. Kolby Smith, Maosheng Zhao, and Steven W. Running, 2012, "Global Bioenergy Capacity as Constrained by Observed Biospheric Productivity Rates," *Bioscience* 62:911–922, for how to convert petagrams of carbon into exajoules.

5. A.D. Barnosky, 2008, "Megafauna Biomass Tradeoff as a Driver of Quaternary and Future Extinctions," *Proceedings of the National Academy of Sciences* 105 (suppl. 1):11543–11548.

6. See ibid. for the nitty-gritty details.

7. The 28.8 percent estimate comes from Helmut Haberl, K. Heinz Erb, Fridolin Krausmann, Veronika Gaube, Alberte Bondeau, Christoph Plutzar, Simone Gingrich, Wolfgang Lucht, and Marina Fischer-Kowalski, 2007, "Quantifying and Mapping the Human Appropriation of Net Primary Production in Earth's Terrestrial Ecosystems," *Proceedings of the National Academy of Sciences* 104:12942–12947. The 40 percent figure is from Peter M. Vitousek, Paul R. Ehrlich, Anne H. Ehrlich, and Pamela A. Matson, 1986, "Human Appropriation of the Products of Photosynthesis," *BioScience* 36:368–373. See also Stuart L. Pimm, 2001, *The World According to Pimm: A Scientist Audits the Earth* (New Brunswick, NJ: Rutgers University Press), for a very readable account of how scientists measure and calculate the amount of Earth's primary productivity that humans appropriate.

8. Keep in mind that oil emissions account for only about 35.2 percent of total emissions from fossil fuels (in 2010). For total emissions from fossil fuels, you also have to add emissions from coal (44.8 percent) and natural gas (20 percent). Data from the U.S. Energy Information Administration, www.eia.gov/cfapps/ipdbproject/iedindex3.cfm.

9. CO_2 release per barrel of oil burned is calculated thus: 5.80 mmbtu/barrel × 20.31 kg C/mmbtu × 44 g CO_2/12 g C × 1 metric ton/1,000 kg = 0.43

metric tons CO_2/barrel; from U.S. Environmental Protection Agency, "Clean Energy," www.epa.gov/cleanenergy/energy-resources/refs.html. The abbreviation "mmbtu" stands for one million British thermal units, or BTUs. A BTU is the amount of heat required to increase the temperature of a pint of water (16 ounces) by one degree Fahrenheit. Take note of the difference between gigatons of CO_2 (abbreviated GtCO2) and gigatons of carbon (abbreviated GtC). Divide gigatons of CO_2 by 3.67 to calculate gigatons of C, or, conversely, multiply gigatons of C by 3.67 to calculate gigatons of CO_2. This difference (and converting back and forth) is important because the relationship of temperature rise to raw emissions (not the emissions that actually end up in the atmosphere, which are much less) is typically expressed in terms of how much adding gigatons of C (not CO_2) increases degrees Celsius.

10. The lower figure (500 gigatons per 1 degree C) is from Steven J. Davis, Long Ca, Ken Caldeira, and Martin I. Hoffert, 2013, "Rethinking Wedges," *Environmental Research Letters* 8(1):011001, doi:10.1088/1748-9326/8/1/011001, which indicates 650 ppm in 2160 assuming the IPCC's A2 scenario and is about the midpoint of the range given in the IPCC's AR5 report; see figure SPM.10 in IPCC, 2013, "Summary for Policymakers," in *Climate Change 2013: The Physical Science Basis,* ed. T.F. Stocker, D. Qin, G.-K. Plattner, M. Tignor, S.K. Allen, J. Boschung, A. Nauels, Y. Xia, V. Bex, and P.M. Midgley (Cambridge: Cambridge University Press), online at www.ipcc.ch/report/ar5/wg1/docs/WGIAR5_SPM_brochure_en.pdf. The higher estimate (600 gigatons per 1 degree rise) is from National Research Council, 2011, *Climate Stabilization Targets: Emissions, Concentrations, and Impacts of Decades to Millennia* (Washington, DC: National Academies Press). These numbers derive from modeling studies such as those detailed in the following: H. Damon Matthews, Nathan P. Gillett, Peter A. Stott, and Kirsten Zickfeld, 2009, "The Proportionality of Global Warming to Cumulative Carbon Emissions," *Nature* 459:833; Malte Meinshausen, Nicolai Meinshausen, William Hare, Sarah C.B. Raper, Katja Frieler, Reto Knutti, David J. Frame, and Myles R. Allen, 2009, "Greenhouse-Gas Emission Targets for Limiting Global Warming to 2°C," *Nature* 458:1158–1162; Myles R. Allen, David J. Frame, Chris Huntingford, Chris D. Jones, Jason A. Lowe, Malte Meinshausen, and Nicolai Meinshausen, 2009, "Warming Caused by Cumulative Carbon Emissions towards the Trillionth Tonne," *Nature* 458:1163–1166.

11. UNDESA, 2011, "World Population Prospects: The 2010 Revision, Vol. II, Demographic Profiles," http://esa.un.org/wpp/Documentation/pdf/WPP2010_Volume-II_Demographic-Profiles.pdf, last accessed February 5, 2014.

12. These calculations assume that we would have dropped to four barrels per person per year by 2012 (which didn't happen) and that number would remain constant into the future, and that population growth would slow so that world population would stabilize at ten billion people by the year 2050.

13. See note 8.

14. According to Steven J. Davis, Long Ca, Ken Caldeira, and Martin I. Hoffert, 2013, "Rethinking Wedges," *Environmental Research Letters* 8(1):011001, doi:10.1088/1748, for the past decade we have been exceeding the emissions projected by the IPCC AR4 A2 scenario. The lower projection (8.6 degrees Fahrenheit, 4.8 degrees Celsius) comes from the IPCC AR5 RCP8.5 scenario. The higher projection (11.5 degrees Fahrenheit, 6.4 degrees Celsius) comes from the IPCC AR4 A1F1 scenario. See note 16 in chapter 3 for a brief explanation of the emissions trajectories specified in each of the IPCC reports. The AR4 report can be found at www.ipcc.ch/pdf/assessment-report/ar4/wg1/ar4-wg1-spm.pdf; the AR5 report at www.ipcc.ch/report/ar5/wg1/docs/WGIAR5_SPM_brochure_en.pdf.

15. Steven J. Davis, Long Ca, Ken Caldeira, and Martin I. Hoffert, 2013, "Rethinking Wedges," *Environmental Research Letters* 8(1):011001, doi:10.1088/1748-9326/8/1/011001.

16. Ibid.

17. Stephen W. Pacala and Robert H. Socolow, 2004, "Stabilization Wedges: Solving the Climate Problem for the Next 50 Years with Current Technologies," *Science* 305:968–972. See also Robert H. Socolow and Stephen W. Pacala, 2006, "A Plan to Keep Carbon in Check," *Scientific American* (September 2006):50–57. See note 9 regarding gigatons of carbon (GtC).

18. I've written extensively about this in another book, Anthony D. Barnosky, 2009, *Heatstroke: Nature in an Age of Global Warming* (Washington, DC: Island Press), which points the way to many articles in the primary scientific literature that document the effects of global climate change so far on the world's species. See also Barnosky, 2013, "Climate Change," in *Grzimek's Animal Life Encyclopedia: Extinction* (Detroit: Gale Publishing), pp. 735–747; and Camille Parmesan, 2006, "Ecological and Evolutionary Responses to Recent Climate Change," *Annual Review of Ecology, Evolution and Systematics* 37:637–639.

19. Steven J. Davis, Long Ca, Ken Caldeira, and Martin I. Hoffert, 2013, "Rethinking Wedges," *Environmental Research Letters* 8(1):011001, doi:10.1088/1748-9326/8/1/011001.

20. PricewaterhouseCoopers LLP, 2012, "Too Late for Two Degrees? Low Carbon Economy Index 2012," www.pwc.com/en_GX/gx/low-carbon-economy-index/assets/pwc-low-carbon-economy-index-2012.pdf:1–16.

21. These two guys are no energy lightweights. Chu was a cowinner of the Nobel Prize in physics in 1997 and was the United States Secretary of Energy during President Barack Obama's first term. Majumdar was right there with him, as Director of the United States Advanced Research Projects Agency–Energy and Acting Under Secretary of Energy. Both rose to scientific acclaim for their work on energy issues at Lawrence Berkeley National Laboratory, and in addition, Majumdar served as an advisor to startup companies and venture capital firms in California's Silicon Valley. Steven Chu and Arun Majumdar, 2012, "Opportunities and Challenges for a Sustainable Energy Future," *Nature* 488:294–303.

22. Regarding buildings, see Philip Farese, 2012, "How to Build a Low-Energy Future," *Nature* 488:275–277.

23. Harold T. Shapiro, Mark S. Wrighton, John F. Ahearne, Allen J. Bard, Jan Beyea, William F. Brinkman, et al., 2009, *America's Energy Future: Technology and Transformation: Summary Edition* (Washington, DC: National Academies Press), p. xii.

24. J. Rogelj, D. L. McCollum, A. Reisinger, M. Meinshausen, and K. Riahi, 2012, "Probabilistic Cost Estimates for Climate Change Mitigation," *Nature* 493:79–83.

25. Steven Chu and Arun Majumdar, 2012, "Opportunities and Challenges for a Sustainable Energy Future," *Nature* 488:294–303.

26. "U.S. Fuel Economy Hits New Record High in February," *Auto Green* blog, http://green.autoblog.com/2012/03/14/u-s-vehicle-fuel-economy-hits-record-in-february/, last accessed February 20, 2013.

27. Steven M. Gorelick, 2010, *Oil Panic and the Global Oil Crisis: Predictions and Myths* (Chichester, UK: Wiley-Blackwell). As of 2008, in Japan and the European Union the standard was 44 miles per gallon; in 2005 and 2010, respectively, China and Australia required 34 miles per gallon.

28. In 2010, transportation accounted for 22 percent of all CO_2 emissions globally, and of that, about 75 percent was related to road travel. This means vehicle emissions produce about 16–17 percent of CO_2 emissions globally—a significant chunk. See International Energy Agency, 2012, "CO_2 Emissions from Fuel Combustion, IEA Statistics, Highlights 2012," www.iea.org/co2highlights/co2highlights.pdf.

29. That number assumes each car on average emitted about 4.8 metric tons of CO_2 per year, as estimated by the U.S. Environmental Protection Agency, "Clean Energy," www.epa.gov/cleanenergy/energy-resources/refs.html. The average of 4.8 tons per car was multiplied by the number of cars 3in 2009, which was 912,536,248, according to World Bank data: http://data.worldbank.org/indicator/IS.VEH.PCAR.P3. Another way to estimate tons of carbon emitted per car per year, which gives a higher number, is to start with 2010 global emissions of CO_2, which were 33.5 gigatons (see http://co2now.org/Current-CO2/CO2-Now/global-carbon-emissions.html and ibid.). Of that, 22 percent (7.37 gigatons) was from the transportation sector, and of the transportation sector, about 75 percent was for road transport, which gives a total of 5.5 gigatons per year for road transport. I used the EPA estimate, but keep in mind the number could be a little higher.

30. This calculation multiplies the number of cars per year by the tons of CO_2 emitted per car per year, then adds the emissions year by year. The starting point is the year 2010, with 923,570,746 cars and 6,892,319,000 people. The average emissions per car (4.8 tons per year) comes from EPA calculations cited in note 29. Number of cars uses the 2009 tally (the 2010 data had not yet been tabulated as of the time I did this) from World Bank data, http://data.worldbank.org/indicator/IS.VEH.PCAR.P3. Number of people in 2010 is from the Population Reference Bureau World Population Data Sheet, www.prb.org/pdf10/10wpds_eng.pdf. The 296 gigatons of CO_2 equates to 81 gigatons of carbon (C) (see note 9).

31. The calculation assumes a 5 percent increase in gas mileage starting in 2014, with a steady increase in mileage until gas mileage is doubled by 2025, after which point it holds steady.

32. World Bank data, http://data.worldbank.org/indicator/IS.VEH.PCAR.P3. The data tables include individual countries for 2010, but the most recent world tabulation is for 2009. The ten countries with the most cars per capita (in 2010) are San Marino (1,139 cars per 1,000 people), Liechtenstein (750), Monaco (732), Luxembourg (665), Iceland (644), the United States (627), Puerto Rico (621), Italy (602), and New Zealand (598). China has 44 cars per 1,000 people. The five countries with the lowest per capita number of cars are Burkina Faso (7 cars per 1,000 people), Myanmar (5), Gambia (5), Bangladesh (2), and Rwanda (2).

33. UNDESA, 2011, "World Population Prospects: The 2010 Revision, Vol. II, Demographic Profiles," http://esa.un.org/wpp/Documentation/pdf/WPP2010_Volume-II_Demographic-Profiles.pdf, last accessed February 5, 2014.

34. This calculation assumes a 1 percent per year increase in the number of cars from 2001 to 2057, then stabilization at the 2057 number (2,138,993,014 cars through 2065).

35. This is the ideal, but if produced using poor practices, biofuels can actually increase rather than decrease the overall carbon footprint. The ultimate carbon footprint of a biofuel depends on the kind of biofuel crop grown, how much it displaces food crops, whether forest has to be cleared to grow the crop or to make up for displaced food crops, and the amount of fossil fuels used (for tractor fuel, fertilizer production, harvest, transport, etc.) to grow and process the biofuel crop. See, for example, Louise Gray, 2010, "Biofuels Cause Four Times More Carbon Emissions," *The Telegraph* (April 22, 2010), www.telegraph.co.uk/earth/environment/climate-change/7614934/Biofuels-cause-four-times-more-carbon-emissions.html; and Damian Carrington, 2012, "Leaked Data: Palm Biodiesel as Dirty as Fuel from Tar Sands," *The Guardian* (January 27, 2012), www.guardian.co.uk/environment/damian-carrington-blog/2012/jan/27/biofuels-biodiesel-ethanol-palm-oil.

36. Gary Roughead, Jeremy Carl, and Manuel Hernandez, 2012, *Powering the Armed Forces: Meeting the Military's Energy Challenges, Report from the Schultz-Stephenson Task Force on Energy Policy* (Stanford, CA: Hoover Institution), www.hoover.org/publications/books/123086.

37. See the news report by Elizabeth Shogren, 2011, "Air Force and Navy Turn to Biofuels," NPR (September 22, 2011), www.npr.org/2011/09/26/140702387/air-force-and-navy-turn-to-bio-fuels; and Tim Worstall, 2012, "Air Force Biofuel at $59 a Gallon: Cheap at Twice the Price," *Forbes* (July 16, 2012), www.forbes.com/sites/timworstall/2012/07/16/air-force-biofuel-at-59-a-gallon-cheap-at-twice-the-price/.

38. Adam Rutherford, 2012, "Synthetic Biology and the Rise of the 'Spider-Goats,'" *The Guardian/Observer* (January 14, 2012), www.guardian.co.uk/science/2012/jan/14/synthetic-biology-spider-goat-genetics.

39. D. Ryan Georgianna and Stephen P. Mayfield, 2012, "Exploiting Diversity and Synthetic Biology for the Production of Algal Biofuels," *Nature* 488:329–335.

40. Pamela P. Peralta-Yahya, Fuzhong Zhang, Stephen B. del Cardayre, and Jay D. Keasling, 2012, "Microbial Engineering for the Production of Advanced Biofuels," *Nature* 488:320–328.

41. Steven Chu and Arun Majumdar, 2012, "Opportunities and Challenges for a Sustainable Energy Future," *Nature* 488:294–303.

42. Nash Kuene, 2012, "Algae: Fuel of the Future?" *National Review Online* (March 8, 2012), www.nationalreview.com/articles/292913/algae-fuel-future-nash-keune.

43. Steven Chu and Arun Majumdar, 2012, "Opportunities and Challenges for a Sustainable Energy Future," *Nature* 488:294–303.

44. Debajyoti Dutta, Debojyoti De, Surabhi Chaudhuri, and Sanjoy K. Bhattacharya, 2005, "Hydrogen Production by Cyanobacteria," *Microbial Cell Factories* 4:36, doi:10.1186/1475-2859-4-36.

45. Paul Stenquist, 2012, "How Green Are Electric Cars? Depends on Where You Plug In," *New York Times,* April 13, 2012, www.nytimes.com/2012/04/15/automobiles/how-green-are-electric-cars-depends-on-where-you-plug-in.html. The basis for the newspaper article was Don Anair and Amine Mahmassani, 2012, "State of Charge: Electric Vehicles' Global Warming Emissions and Fuel-Cost Savings across the United States," Union of Concerned Scientists, www.ucsusa.org/assets/documents/clean_vehicles/electric-car-global-warming-emissions-report.pdf.

46. U.S. Energy Information Administration, "Frequently Asked Questions," www.eia.gov/tools/faqs. One Btu is equivalent to about 1,055 joules, or the amount of energy it takes to heat one pound of water by one degree Fahrenheit.

47. Based on population growth to ten million by 2050, then stabilization at that number, and per capita emissions from coal use as of 2010, which was about 14.23 gigatons of CO_2, or 2.06 tons per person. Data from the U.S. Energy Information Administration, www.eia.gov/cfapps/ipdbproject/iedindex3.cfm.

48. Mark Fulton, Nils Mellquist, Saya Katasel, and Joel Bluestein, 2011, "Comparing Life-Cycle Greenhouse Gas Emissions from Natural Gas and Coal," Worldwatch Institute, www.worldwatch.org/system/files/pdf/Natural_Gas_LCA_Update_082511.pdf.

49. Basically, fracking fractures the rock underground, which allows otherwise trapped gas and oil to flow into collection wells.

50. Steven Chu and Arun Majumdar, 2012, "Opportunities and Challenges for a Sustainable Energy Future," *Nature* 488:294–303.

51. Ibid.

52. Christian Azar, Daniel J.A. Johansson, and Niclas Mattsson, 2013, "Meeting Global Temperature Targets: The Role of Bioenergy with Carbon Capture and Storage," *Environmental Research Letters* 8(3):034004, doi:10.1088/1748-9326/8/3/034004.

53. Mark Z. Jacobson and Cristina L. Archer, 2012, "Saturation Wind Power Potential and Its Implications for Wind Energy," *Proceedings of the National Academy of Sciences* 109:15679–15684.

54. Steven Chu and Arun Majumdar, 2012, "Opportunities and Challenges for a Sustainable Energy Future," *Nature* 488:294–303.

55. U.S. Environmental Protection Agency, 2013, "Ocean Energy," www.epa.gov/region1/eco/energy/re_ocean.html, accessed February 20, 2013.

56. Solarbuzz, 2010, "Solar Energy Market Growth," www.solarbuzz.com/facts-and-figures/markets-growth/market-growth; "Wind Power: Emerging Markets Drive Growth," *Reinforced Plastics* (2012) 56:32–33.

57. Mark Z. Jacobson and Mark A. Delucchi, 2009, "A Path to Sustainable Energy by 2030," *Scientific American* (November 2009):58–65. For a more technical treatment, see Mark Z. Jacobson and Mark A. Delucchi, 2011, "Providing All Global Energy with Wind, Water, and Solar Power, Part I: Technologies, Energy Resources, Quantities and Areas of Infrastructure, and Materials," *Energy Policy* 39:1154–1169; and Delucchi and Jacobson, 2011, "Providing All Global Energy with Wind, Water, and Solar Power, Part II: Reliability, System and Transmission Costs, and Policies," *Energy Policy* 39:1170–1190.

58. Regarding wave converters, since wind causes the waves, you could classify this as wind-generated energy, as Jacobson and Delucchi do.

59. See the citations in note 57 for more discussion of points summarized in this paragraph. Regarding advances in battery technology, also see "Better Batteries: New Technology Improves Both Energy Capacity and Charge Rate in Rechargeable Batteries," *Science Daily* (November 17, 2011), www.sciencedaily.com/releases/2011/11/111114142047.htm; and the technical article by Xin Zhao, Cary M. Hayner, Mayfair C. Kung, and Harold H. Kung, 2011, "In-Plane Vacancy-Enabled High-Power Si-Graphene Composite Electrode for Lithium-Ion Batteries," *Advanced Energy Materials* 1(6):1079, doi:10.1002/aenm.201100426.

60. For an excellent treatment of the peak oil debate, see Steven M. Gorelick, 2010, *Oil Panic and the Global Oil Crisis: Predictions and Myths* (Chichester, UK: Wiley-Blackwell). The upshot is that we are in no danger of running out of economically viable oil within the next century, especially given new technology that makes extraction of oil from tar sands possible within needed profit margins. Therefore it is unrealistic to depend on market forces alone to stimulate a shift from fossil fuels to a carbon-neutral energy landscape.

61. Steven Gorelick (ibid.) recounts this, and Steven Chu and Arun Majumdar also quote it in their paper (Chu and Majumdar, 2012, "Opportunities and Challenges for a Sustainable Energy Future," *Nature* 488:294–303).

62. See Steven Chu and Arun Majumdar, 2012, "Opportunities and Challenges for a Sustainable Energy Future," *Nature* 488:294–303; Mark Z. Jacobson and Mark A. Delucchi, 2009, "A Path to Sustainable Energy by 2030," *Scientific American* (November 2009):58–65; and Mark A. Delucchi and Mark Z. Jacobson, 2011, "Providing All Global Energy with Wind, Water, and Solar Power, Part II: Reliability, System and Transmission Costs, and Policies," *Energy Policy* 39:1170–1190.

63. See Michelle Nijhuis, 2008, "What's Killing the Aspen?" *Smithsonian Magazine,* www.smithsonianmag.com/science-nature/Phenomena-Rocky-Aspens-200812.html; and William R. L. Anderegg, Joseph A. Berry, Duncan D. Smith, John S. Sperry, Leander D. L. Anderegg, and Christopher B. Field, 2012, "The Roles of Hydraulic and Carbon Stress in a Widespread Climate-Induced Forest Die-Off," *Proceedings of the National Academy of Sciences* 109:233–237.

5. FOOD

1. One eighteen-incher was enough for breakfast. After that was assured, it turned into full-fledged catch-and-release sport fishing.

2. A quote from a tuna fisherman, reported in Paul Greenberg, 2011, *Four Fish: The Future of the Last Wild Food* (New York: Penguin), p. 205.

3. "Ice-free land" excludes Antarctica, high mountains that still have permanent glaciers, and ice-covered parts of the Arctic and Greenland. The estimate for ice-free land comes from Jonathan A. Foley, Navin Ramankutty, Kate A. Brauman, Emily S. Cassidy, James S. Gerber, Matt Johnston, et al., 2011, "Solutions for a Cultivated Planet," *Nature* 478:337–342; and Jonathan A. Foley, 2011, "Can We Feed the Planet?" *Scientific American* (November 2011):60–65. The estimate for percentage of total land surface uses numbers given in Foley et al. for hectares of crop and pastureland and assumes the total land area of Earth (glaciated regions included) is about 150,000,000 km^2. Estimates of total land area on Earth vary from about 148,000,000 to 153,000,000 km^2. About 10.4 percent of the land surface is estimated to be covered with ice.

4. P. S. Martin and C. Szuter, 1999, "War Zones and Game Sinks in Lewis and Clark's West," *Conservation Biology* 13(1):36–45.

5. Kim Murphy, 2013, "More Than 550 Wolves Taken by Hunters and Trappers in Rockies," *Los Angeles Times* (March 6, 2013), http://articles.latimes.com/2013/mar/06/nation/la-na-nn-wolves-idaho-montana-hunt-trap-20130305; *Science Daily*, 2004, "Lion Attacks on Livestock in Africa Are Significant but Manageable," www.sciencedaily.com/releases/2004/03/040319073054.htm,

reporting on the study by Bruce D. Patterson, Samuel M. Kasiki, Edwin Selempo, and Roland W. Kays, 2004, "Livestock Predation by Lions (*Panthera leo*) and Other Carnivores on Ranches Neighboring Tsavo National Park, Kenya," *Biological Conservation* 119:507–516; James Munyeki, 2013, "Disaster as Jumbos Destroy Crops," *Kenya Standard* (April 4, 2013), https://www.standardmedia.co.ke/?articleID=2000080888&story_title=Kenya--Disaster-as-jumbos-destroy-crops.

6. James A. Estes, John Terborgh, Justin S. Brashares, Mary E. Power, Joel Berger, William J. Bond, et al., 2011, "Trophic Downgrading of Planet Earth," *Science* 333:301–306.

7. The same principles hold true in aquatic systems. Kelp forests off the coast of California, for example, are maintained because sea otters eat the sea urchins that in turn eat the kelp: no sea otters, too many sea urchins, no kelp. In Michigan lakes, large-mouth bass keep water clear by eating fish that would otherwise eat zooplankton. The zooplankton thus thrive in numbers that allow them to outcompete phytoplankton, which in turn promotes water clarity and allows many other species to survive. See ibid.

8. Many studies suggest somewhere in the vicinity of 5,000 to 7,000 individuals as a minimum viable population size. See thorough discussions in Curtis H. Flather, Gregory D. Hayward, Steven R. Beissinger, and Philip A. Stephens, 2011, "Minimum Viable Populations: Is There a 'Magic Number' for Conservation Practitioners?" *Trends in Ecology and Evolution* 26:307–316; and David H. Reeda, Julian J. O'Grady, Barry W. Brook, Jonathan D. Ballou, and Richard Frankham, 2003, "Estimates of Minimum Viable Population Sizes for Vertebrates and Factors Influencing Those Estimates," *Biological Conservation* 113:23–34.

9. Richard Levins, 1969, "Some Demographic and Genetic Consequences of Environmental Heterogeneity for Biological Control," *Bulletin of the Entomological Society of America*, 15:237–240. See also Ilka Haanski and Michael Gilpin, 1991, "Metapopulation Dynamics: Brief History and Conceptual Domain," *Biological Journal of the Linnaean Society* 42:3–16.

10. International Union for the Conservation of Nature, 2013, "*Tapirus indicus*," www.iucnredlist.org/details/21472/0.

11. W.H. Schlesinger, J.F. Reynolds, G.L. Cunningham, L.F. Huenneke, W.M. Jarrell, R.A. Virginia, and W.G. Whitford, 1990, "Biological Feedbacks in Global Desertification," *Science* 147:1043–1048.

12. Robert Stewart, 2010, "Desertification in the Sahel," in *Environmental Science in the 21st Century, an Online Text Book*, http://oceanworld.tamu.edu/resources/environment-book/desertificationinsahel.html.

13. Paolo D'Odorico, Abinash Bhattachan, Kyle F. Davis, Sujith Ravi, and Christiane W. Runyan, 2013, "Global Desertification: Drivers and Feedbacks," *Advances in Water Resources* 51:326–344, doi:10.1016/j.advwatres.2012.01.013.

14. Chad R. Reid, Sherel Goodrich, and James E. Bowns, 2008, "Cheatgrass and Red Brome: History and Biology of Two Invaders," *USDA Forest Service Proceedings* RMRS-P-52:27–32.

15. Kaitlin Maguire, at the time a graduate student working on her PhD project on the paleobiology of mammals in the John Day region of Oregon, mentioned that a rancher had told her this. Neither she nor I have been able to find written verification.

16. BASF Corporation, "Non-Native Cheatgrass Cheats Farmers and Ranchers Out of Their Land," promotional brochure, bettervm.basf.us_frequently-asked-questions_literature_cheatgrass-technical-bulletin-.pdf.

17. For effects on fire frequency, see Jennifer K. Balch, Bethany A. Bradley, Carla M. D'Antonio, and José Gómez-Dans, 2013, "Introduced Annual Grass Increases Regional Fire Activity across the Arid Western USA (1980–2009)," *Global Change Biology* 19:173–183. For effects on biodiversity, see Lucas K. Hall, 2012, "Effect of Cheatgrass on Abundance of the North American Deermouse (*Peromyscus maniculatus*)," *Southwestern Naturalist* 57(2):166–169; and Rebecca C. Terry, Cheng (Lily) Li, and Elizabeth A. Hadly, 2011, "Predicting Small-Mammal Responses to Climatic Warming: Autecology, Geographic Range, and the Holocene Fossil Record," *Global Change Biology* 17:3019–3034.

18. Fabian Menalled, Jane Mangold, and Ed Davis, 2008, "Cheatgrass: Identification, Biology and Integrated Management," *Montana State University Extension Montguide* MT200811AG, http://ipm.montana.edu/cropweeds/montguides/cheatgrass.pdf.

19. See Rebecca C. Terry, Cheng (Lily) Li, and Elizabeth A. Hadly, 2011, "Predicting Small-Mammal Responses to Climatic Warming: Autecology, Geographic Range, and the Holocene Fossil Record," *Global Change Biology* 17:3019–3034; and Lucas K. Hall, 2012, "Effect of Cheatgrass on Abundance of the North American Deermouse (*Peromyscus maniculatus*)," *Southwestern Naturalist* 57(2):166–169.

20. International Union for the Conservation of Nature, 2013, "Why Is Biodiversity in Crisis?" www.iucn.org/iyb/about/biodiversity_crisis/.

21. Helmut J. Geist and Eric F. Lambin, 2002, "Proximate Causes and Underlying Driving Forces of Tropical Deforestation," *Bioscience* 52:143–150.

22. Estimates of the amount of rainforest left in the world range from less than 8 million square kilometers (almost 2 billion acres) to between 6.2 and 6.8

million square kilometers (about 1.5 to 1.7 billion acres), with the latter numbers equating to about 5 percent of the ice free land area. See www.nature.org/ourinitiatives/urgentissues/rainforests/rainforests-facts.xml and http://rainforests.mongabay.com/0101.htm.

23. For the increase in agricultural land, see Erle Ellis, 2011, "Anthropogenic Transformation of the Terrestrial Biosphere," *Philosophical Transactions of the Royal Society* A 369:1010–1035.

24. N.V. Fedoroff, D.S. Battisti, R.N. Beachy, P.J.M. Cooper, D.A. Fischhoff, C.N. Hodges, et al., 2010, "Radically Rethinking Agriculture for the 21st Century," *Science* 327:833–834; David B. Lobell, Kenneth G. Cassman, and Christopher B. Field, 2009, "Crop Yield Gaps: Their Importance, Magnitudes, and Causes," *Annual Review of Environment and Resources* 34:179–204; James N. Galloway, Marshall Burke, G. Eric Bradford, Rosamond Naylor, Walter Falcon, Ashok K. Chapagain, et al., 2007, "International Trade in Meat: The Tip of the Pork Chop," *Ambio* 36:622–629; Rosamond Naylor, Henning Steinfeld, Walter Falcon, James Galloway, Vaclav Smil, Eric Bradford, Jackie Alder, and Harold Mooney, 2005, "Losing the Links between Livestock and Land," *Science* 310:1621–1622. See also the work and programs of the Consultative Group on International Agricultural Research, www.cgiar.org/.

25. Jonathan A. Foley, Navin Ramankutty, Kate A. Brauman, Emily S. Cassidy, James S. Gerber, Matt Johnston, et al., 2011, "Solutions for a Cultivated Planet," *Nature* 478:337–342. See also Jonathan Foley, 2011, "Can We Feed the Planet?" *Scientific American* (November 2011):60–65.

26. See Jonathan A. Foley, Navin Ramankutty, Kate A. Brauman, Emily S. Cassidy, James S. Gerber, Matt Johnston, et al., 2011, "Solutions for a Cultivated Planet," *Nature* 478:337–342, and David B. Lobell, Kenneth G. Cassman, and Christopher B. Field, 2009, "Crop Yield Gaps: Their Importance, Magnitudes, and Causes," *Annual Review of Environment and Resources* 34:179–204.

27. Lewis H. Ziska, James A. Bunce, Hiroyuki Shimono, David R. Gealy, Jeffrey T. Baker, Paul C.D. Newton, et al., 2012, "Food Security and Climate Change: On the Potential to Adapt Global Crop Production by Active Selection to Rising Atmospheric Carbon Dioxide," *Proceedings of the Royal Society* B 279:4097–4105, www.ncbi.nlm.nih.gov/pmc/articles/PMC3441068/pdf/rspb20121005.pdf.

28. David B. Lobell, Marshall B. Burke, Claudia Tebaldi, Michael D. Mastrandrea, Walter P. Falcon, and Rosamond L. Naylor, 2008, "Prioritizing Climate Change Adaptation Needs for Food Security in 2030," *Science* 319:607–610. See also C.L. Walthall, J. Hatfield, P. Backlund, L. Lengnick, E. Marshall,

M. Walsh, et al., 2012, "Climate Change and Agriculture in the United States: Effects and Adaptation," *USDA Technical Bulletin* 1935, www.usda.gov/oce/climate_change/effects_2012/CC%20and%20Agriculture%20Report%2802-04-2013%29b.pdf.

29. Sandra L. Postel, 2012, "Drip Irrigation Expanding Worldwide," *National Geographic Newswatch: Water Currents,* http://newswatch.nationalgeographic.com/2012/06/25/drip-irrigation-expanding-worldwide/.

30. Sandra L. Postel, 2012, "Getting More Crop per Drop," in Linda Starke, ed., *State of the World: Innovations That Nourish the Planet* (New York: W. W. Norton), pp. 39–48, www.worldwatch.org/sow11. See also Jonathan A. Foley, Navin Ramankutty, Kate A. Brauman, Emily S. Cassidy, James S. Gerber, Matt Johnston, et al., 2011, "Solutions for a Cultivated Planet," *Nature* 478:337–342.

31. Robert I. McDonald, Pamela Green, Deborah Balk, Balazs M. Fekete, Carmen Revenga, Megan Todd, and Mark Montgomery, 2011, "Urban Growth, Climate Change, and Freshwater Availability," *Proceedings of the National Academy of Sciences* 108:6312–6317; Jeff Smith, 2012, "Growing Food Demand Strains Energy, Water Supplies," *National Geographic News* (April 6, 2012), http://news.nationalgeographic.com/news/energy/2012/04/120406-food-water-energy-nexus/; Marianne Lavelle and Thomas K. Grose, 2013, "Water Demand for Energy to Double by 2035," *National Geographic News* (January 30, 2013), http://news.nationalgeographic.com/news/energy/2013/01/130130-water-demand-for-energy-to-double-by-2035/.

32. Frank A. Ward and Manuel Pulido-Velazquez, 2008, "Water Conservation in Irrigation Can Increase Water Use," *Proceedings of the National Academy of Sciences* 105:18215–18220.

33. P. M. Vitousek, R. Naylor, T. Crews, M. B. David, L. E. Drinkwater, E. Holland, et al., 2009, "Nutrient Imbalances in Agricultural Development," *Science* 324:1519–1520.

34. See Jonathan A. Foley, Navin Ramankutty, Kate A. Brauman, Emily S. Cassidy, James S. Gerber, Matt Johnston, et al., 2011, "Solutions for a Cultivated Planet," *Nature* 478:337–342.

35. These numbers are derived by using the estimate of Foley et al. (ibid.) of 3,500,000 square kilometers for the amount of cropland devoted to animal feed, and the estimate of about 4,000,000 square kilometers of cropland devoted to biofuel crops. The total of animal-feed plus biofuel acreage therefore is about 7,500,000 km^2. Foley et al. calculate that converting all animal-feed and biofuel acreage to crops for human consumption would increase caloric yield by 49 percent and food production by 28 percent. Because the

present animal-feed acreage represents about 47 percent of the total of animal-feed plus biofuel cropland, multiplying 49 percent and 28 percent, respectively, by 0.47 gives the increase expected (23 percent and 13 percent, respectively) from converting animal-feed land to crops for human consumption. In a like manner, the respective increases from converting biofuel lands to food-crop lands is obtained by multiplying 49 percent × 0.53 (the percentage of proportion of non-food-producing croplands devoted to biofuel crops). The estimates for square kilometers of land devoted to biofuel production come from the following sources: W. Kolby Smith, Maosheng Zhao, and Steven W. Running, 2012, "Global Bioenergy Capacity as Constrained by Observed Biospheric Productivity Rates," *Bioscience* 62:911–922, give a range of 3.9 to 4.7 million square kilometers presently devoted to biofuels. Régis Rathmann, Alexandre Szklo, and Roberto Schaeffer, 2010, "Land Use Competition for Production of Food and Liquid Biofuels: An Analysis of the Arguments in the Current Debate," *Renewable Energy* 35:14–22, estimate biofuel croplands to be 2.85 percent of arable land. Assuming the Foley et al. estimate of total cropland roughly represents Rathman et al.'s "arable land," multiplying 15,300,000 km^2 by 0.285 gives 4,360,500 km^2.

36. Vaclav Smil, 2010, "Improving Efficiency and Reducing Waste in Our Food System," *Environmental Sciences* 1(1):17–26.

37. Jonathan A. Foley, Navin Ramankutty, Kate A. Brauman, Emily S. Cassidy, James S. Gerber, Matt Johnston, et al., 2011, "Solutions for a Cultivated Planet," *Nature* 478:337–342.

38. Vaclav Smil, 2010, "Improving Efficiency and Reducing Waste in Our Food System," *Environmental Sciences* 1(1):17–26.

39. Paul L. Koch and Anthony D. Barnosky, 2006, "Late Quaternary Extinctions: State of the Debate," *Annual Review of Ecology, Evolution, and Systematics* 37:215–250.

40. See, for example, Wildlife Conservation Society, 2012, "Bushmeat Pushes African Species to the Brink," *Science Daily* (October 25, 2012), www.sciencedaily.com/releases/2012/10/121025140702.htm, accessed April 3, 2013; and Richard K.B. Jenkins, Aidan Keane, Andrinajoro R. Rakotoarivelo, Victor Rakotomboavonjy, Felicien H. Randrianandrianina, H. Julie Razafimanahaka, Sylvain R. Ralaiarimalala, and Julia P.G. Jones, 2011, "Analysis of Patterns of Bushmeat Consumption Reveals Extensive Exploitation of Protected Species in Eastern Madagascar," *PLoS ONE* 6(12):e27570, doi:10.1371/journal.pone.0027570.

41. P. Lindsey, G. Balme, M. Becker, C. Begg, C. Bento, C. Bocchino, et al., 2013, "Illegal Hunting and the Bushmeat Trade in Savanna Africa: Drivers,

Impacts and Solutions to Address the Problem," Panthera / Zoological Society of London / Wildlife Conservation Society Report, www.panthera.org/sites/default/files/bushmeat%20report%20v2%20lo_o.pdf, accessed April 3, 2013; Justin S. Brashares, Peter Arcese, Moses K. Sam, Peter B. Coppolillo, A.R.E. Sinclair, and Andrew Balmford, 2004, "Bushmeat Hunting, Wildlife Declines, and Fish Supply in West Africa," *Science* 306:1180–1183.

42. P. Lindsey, G. Balme, M. Becker, C. Begg, C. Bento, C. Bocchino, et al., 2013, "Illegal Hunting and the Bushmeat Trade in Savanna Africa: Drivers, Impacts and Solutions to Address the Problem," Panthera / Zoological Society of London / Wildlife Conservation Society Report, www.panthera.org/sites/default/files/bushmeat%20report%20v2%20lo_o.pdf, accessed April 3, 2013.

43. Jonathan M. Hoekstra, Jennifer L. Molnar, Michael Jennings, Carmen Revenga, Mark D. Spaulding, Timothy M. Boucher, James C. Robertson, Thomas J. Heibel, and Katherine Ellison, 2010, *The Atlas of Global Conservation* (Berkeley: University of California Press).

44. Robert J. Diaz and Rutger Rosenberg, 2008, "Spreading Dead Zones and Consequences for Marine Ecosystems," *Science* 321:926–929.

45. Jonathan M. Hoekstra, Jennifer L. Molnar, Michael Jennings, Carmen Revenga, Mark D. Spaulding, Timothy M. Boucher, James C. Robertson, Thomas J. Heibel, and Katherine Ellison, 2010, *The Atlas of Global Conservation* (Berkeley: University of California Press).

46. The IUCN, as of 2008, listed 32 of the 115 cetacean species as either endangered or critically endangered (cetaceans include whales, dolphins, and porpoises). See http://cmsdata.iucn.org/downloads/cetacean_table_for_website.pdf.

47. Paul Greenberg, 2011, *Four Fish: The Future of the Last Wild Food* (New York: Penguin), p. 35.

48. Jan Schipper, Janice S. Chanson, Federica Chiozza, Neil A. Cox, Michael Hoffmann, Vineet Katariya, et al., 2008, "The Status of the World's Land and Marine Mammals: Diversity, Threat, and Knowledge," *Science* 322:225–230.

49. IUCN, 2008, www.iucnredlist.org/details/3590/0; www.iucnredlist.org/details/14549/0; and www.iucnredlist.org/details/44187/0.

50. Rebecca L. Lewison, Sloan A. Freeman, and Larry B. Crowder, 2004, "Quantifying the Effects of Fisheries on Threatened Species: The Impact of Pelagic Longlines on Loggerhead and Leatherback Sea Turtles," *Ecology Letters* 7:221–231.

51. In fact, predator-prey regulation is a little more complex than this simple two-species model. In the case of the Canadian lynx and snowshoe hares,

food availability for the hares also enters into the equation. Nevertheless, the net effect of the out-of-phase, boom-and-bust regulation of the two species by each other is the same. See Nils Christian Stenseth, Wilhelm Falck, Ottar N. Bjørnstad, and Charles J. Krebs, 1997, "Population Regulation in Snowshoe Hare and Canadian Lynx: Asymmetric Food Web Configurations between Hare and Lynx," *Proceedings of the National Academy of Sciences* 94:5147–5152.

52. Paul L. Koch and Anthony D. Barnosky, 2006, "Late Quaternary Extinctions: State of the Debate," *Annual Review of Ecology, Evolution, and Systematics* 37:215–250.

53. B.B. Collette, K.E. Carpenter, B.A. Polidoro, M.J. Juan-Jordá, A. Boustany, D. Die, et al., 2011, "High Value and Long Life: Double Jeopardy for Tunas and Billfishes," *Science* 333:291–292, report that current adult biomass is about 5 percent of estimated original biomass. In 2011, the IUCN reported an "estimated 85.4% decline in spawning stock biomass over the past 36 years from 1973 to 2009," www.iucnredlist.org/details/21858/0.

54. IUCN, 2011, www.iucnredlist.org/details/21860/0.

55. Pacific Bluefin Working Group, 2013, "Stock Assessment of Pacific Bluefin Tuna in 2012," http://isc.ac.affrc.go.jp/pdf/Stock_assessment/Stock%20Assessment%20of%20Pacific%20Bluefin%20Assmt%20Report%20-%20May15.pdf. For the 96.4 percent figure, see Bryan Walsh, 2013, "The Pacific Bluefin Tuna Is Going, Going . . . ," *Time Magazine* (January 11, 2013), http://science.time.com/2013/01/11/the-pacific-bluefin-tuna-is-almost-gone/.

56. *Thunnus obesus:* IUCN, 2011, http://www.iucnredlist.org/details/21859/0. *Thunnus alalunga:* IUCN, 2011, http://www.iucnredlist.org/details/21856/0.

57. IUCN, 2011, http://www.iucnredlist.org/details/21857/0.

58. Rosamond L. Naylor, Rebecca J. Goldburg, Jurgenne H. Primavera, Nils Kautsky, Malcolm C.N. Beveridge, Jason Clay, Carl Folke, Jane Lubchenco, Harold Mooney, and Max Troell, 2000, "Effect of Aquaculture on World Fish Supplies," *Nature* 405:1017–1024.

59. Paul Greenberg, 2011, *Four Fish: The Future of the Last Wild Food* (New York: Penguin).

60. I've written in more detail about this elsewhere, as have many other people. See, for example, chapter 13, "Bad Company," in Anthony D. Barnosky, 2009, *Heatstroke: Nature in an Age of Global Warming* (Washington, DC: Island Press), and citations therein.

61. Paul Rogers, 2006, "Economy of Scales," *Stanford Magazine* (March 20, 2006), http://foodsecurity.stanford.edu/news/aquaculture_specialist_rosamond_

naylor_explores_whether_fish_farms_can_sustainably_meet_the_growing_world_demand_20060320.

62. Ibid. You might argue that the same three-to-one ratio applies to wild salmon, and you would be right. The problem is that by farming salmon, we are keeping their numbers artificially high, thus overriding the natural balance between predator and prey that exists in nature absent human interference.

63. Rosamond L. Naylor, Rebecca J. Goldburg, Jurgenne H. Primavera, Nils Kautsky, Malcolm C.N. Beveridge, Jason Clay, Carl Folke, Jane Lubchenco, Harold Mooney, and Max Troell, 2000, "Effect of Aquaculture on World Fish Supplies," *Nature* 405:1017–1024.

64. Rosamond L. Naylor, Ronald W. Hardy, Dominique P. Bureau, Alice Chiu, Matthew Elliott, Anthony P. Farrell, et al., 2009, "Feeding Aquaculture in an Era of Finite Resources," *Proceedings of the National Academy of Sciences* 106:15103–15110.

65. Paul Rogers, 2006, "Economy of Scales," *Stanford Magazine* (March 20, 2006), http://foodsecurity.stanford.edu/news/aquaculture_specialist_rosamond_naylor_explores_whether_fish_farms_can_sustainably_meet_the_growing_world_demand_20060320.

66. J.P. Volpe, J. Gee, M. Beck, and V. Ethier, 2011, "How Green Is Your Eco-Label? Comparing the Environmental Benefits of Marine Aquaculture Standards," http://web.uvic.ca/~serg/papers/GAPI_Benchmarking_Report_2011.pdf.

67. David Perlman, 2013, "Salmon Raised in a Rice Field Thrive," *San Francisco Chronicle* (April 4, 2013), www.sfchronicle.com/science/article/Salmon-raised-in-a-rice-field-thrive-4408232.php.

68. Elisabeth Rosenthal, 2011, "Another Side of Tilapia, the Perfect Factory Fish," *New York Times* (May 2, 2011), www.nytimes.com/2011/05/02/science/earth/02tilapia.html.

69. Paul Greenburg, 2011, *Four Fish: The Future of the Last Wild Food* (New York: Penguin).

70. Rosamond L. Naylor, Rebecca J. Goldburg, Jurgenne H. Primavera, Nils Kautsky, Malcolm C.N. Beveridge, Jason Clay, Carl Folke, Jane Lubchenco, Harold Mooney, and Max Troell, 2000, "Effect of Aquaculture on World Fish Supplies," *Nature* 405:1017–1024.

71. B.B. Collette, K.E. Carpenter, B.A. Polidoro, M.J. Juan-Jordá, A. Boustany, D.J. Die, et al., 2011, "High Value and Long Life: Double Jeopardy for Tunas and Billfishes," *Science* 333:291–292.

72. Paul Greenberg, 2011, *Four Fish: The Future of the Last Wild Food* (New York: Penguin).

73. Monterey Bay Aquarium Seafood Watch, www.montereybayaquarium .org/cr/seafoodwatch.aspx.

74. Jacob H. Lowenstein, George Amato, and Sergios-Orestis Kolokotronis, 2009, "The Real *maccoyii:* Identifying Tuna Sushi with DNA Barcodes—Contrasting Characteristic Attributes and Genetic Distances," *PLoS ONE* 4(11):e7866, doi:10.1371/journal.pone.0007866.

75. Ibid., p. 7.

76. Bryan Walsh, 2013, "The Pacific Bluefin Tuna Is Going, Going … ," *Time Magazine* (January 11, 2013), http://science.time.com/2013/01/11/the-pacific-bluefin-tuna-is-almost-gone/. Walsh's report is based on Pacific Bluefin Working Group, 2013, "Stock Assessment of Pacific Bluefin Tuna in 2012," http://isc.ac.affrc.go.jp/pdf/Stock_assessment/Stock%20Assessment%20of%20Pacific%20Bluefin%20Assmt%20Report%20-%20May15.pdf.

6. MONEY

1. Jeffrey Gettleman, 2012, "Elephants Dying in Epic Frenzy as Ivory Fuels Wars and Profits," *New York Times* (September 3, 2012).

2. Gretchen C. Daily, Tore Söderqvist, Sara Aniyar, Kenneth Arrow, Partha Dasgupta, Paul R. Ehrlich, et al., 2000, "The Value of Nature and the Nature of Value," *Science* 289:395–396.

3. The CIA Factbook (https://www.cia.gov/library/publications/the-world-factbook/geos/ke.html) lists Kenya's 2010 GDP at $69.33 billion. USAID considers tourism, "for which the major draw is wildlife," to account for 14 percent of total GDP (http://kenya.usaid.gov/programs/environment).

4. Anthony D. Barnosky, James H. Brown, Gretchen C. Daily, Rodolfo Dirzo, Anne H. Ehrlich, Paul R. Ehrlich, et al., 2014, "Introducing the *Scientific Consensus on Maintaining Humanity's Life Support Systems in the 21st Century: Information for Policy Makers,*" *Anthropocene Review* 1:78–109, doi:10.1177/2053019613516290, with background information and foreign-language versions available at http://consensusforaction.stanford.edu/. See also James W.C. White, Richard B. Alley, David Archer, Anthony D. Barnosky, Jonathan Foley, Rong Fu, et al., 2013, *Abrupt Impacts of Climate Change: Anticipating Surprises* (Washington, DC: National Academies Press).

5. Kirsten Dow and Taylor Downing, 2013, *The Atlas of Climate Change* (Berkeley: University of California Press), pp. 1–112.

6. William Von Hippel, Frank A. Von Hippel, Norman Chan, and Clara Cheng, 2005, "Exploring the Use of Viagra in Place of Animal and Plant Potency Products in Traditional Chinese Medicine," *Environmental Conservation* 32(3):235–238.

7. Richard Ellis, 2005, *Tiger Bone and Rhino Horn: The Destruction of Wildlife for Traditional Chinese Medicine* (Washington, DC: Island Press). This book gives an authoritative account of the use of tigers, rhinos, and other species in Chinese traditional medicine. See also "Trade in Tiger Parts," www.singlevisioninc.org/tiger_body_parts_sold.html, last accessed June 4, 2013. This website lists herbal or other traditional medicines that can be substituted for purported tiger cures in the absence of modern pharmaceuticals.

8. Yu Tian, Jianguo Wu, Andrew T. Smith, Tianming Wang, Xiaojun Kou, and Jianping Ge, 2011, "Population Viability of the Siberian Tiger in a Changing Landscape: Going, Going and Gone?" *Ecological Modelling* 222:3166–3180. See also John S. Kenney, James L. D. Smith, Anthony M. Starfield, and Charles W. McDougal, 1995, "The Long-Term Effects of Tiger Poaching on Population Viability," *Conservation Biology* 9(5):1127–1133.

9. Richard Ellis, 2005, *Tiger Bone and Rhino Horn: The Destruction of Wildlife for Traditional Chinese Medicine* (Washington, DC: Island Press).

10. Kristen Nowell on behalf of TRAFFIC, 2012, "Species Trade and Conservation: Rhinoceroses: Assessment of Rhino Horn as a Traditional Medicine," report prepared for the CITES Secretariat, SC62 Doc. 47.2 Annex (Rev. 2), April 2012, www.cites.org/eng/com/sc/62/E62-47-02-A.pdf.

11. See Richard Ellis, 2005, *Tiger Bone and Rhino Horn: The Destruction of Wildlife for Traditional Chinese Medicine* (Washington, DC: Island Press).

12. Frank Langfitt, 2013, "Vietnam's Appetite for Rhino Horn Drives Poaching in Africa," NPR News (May 13, 2013), www.npr.org/blogs/parallels/2013/05/14/181587969/Vietnams-Appetite-For-Rhino-Horn-Drives-Poaching-In-Africa; and Gwynn Guilford, 2013, "Why Does a Rhino Horn Cost $300,000? Because Vietnam Thinks It Cures Cancer and Hangovers," *The Atlantic* (May 15, 2013), www.theatlantic.com/business/archive/2013/05/why-does-a-rhino-horn-cost-300-000-because-vietnam-thinks-it-cures-cancer-and-hangovers/275881/.

13. As reported in Richard Ellis, 2005, *Tiger Bone and Rhino Horn: The Destruction of Wildlife for Traditional Chinese Medicine* (Washington, DC: Island Press).

14. Ibid.

15. Frank Langfitt, 2013, "Vietnam's Appetite for Rhino Horn Drives Poaching in Africa," NPR News (May 13, 2013), www.npr.org/blogs/parallels/2013/05/14/181587969/Vietnams-Appetite-For-Rhino-Horn-Drives-Poaching-In-Africa.

16. Gwynn Guilford, 2013, "Why Does a Rhino Horn Cost $300,000? Because Vietnam Thinks It Cures Cancer and Hangovers," *The Atlantic* (May 15, 2013), www.theatlantic.com/business/archive/2013/05/why-does-a-rhino-horn-cost-300-000-because-vietnam-thinks-it-cures-cancer-and-hangovers/275881/.

17. IUCN Red List, 2012, *Ceratotherium simum,* www.iucnredlist.org/details/4185/0.

18. Annimiticus, "South Africa: 188 Rhinos Killed in 85 Days," http://annamiticus.com/tag/rhinos-killed-in-south-africa-2013/, accessed June 7, 2013.

19. Annimiticus, "South Africa: 313 Rhinos Killed in 134 Days," http://annamiticus.com/2013/05/16/south-africa-313-rhinos-killed-in-134-days/, accessed June 7, 2013; Adam Vaughan, 2014, "More than 1,000 Rhinos Killed in South Africa in 2013," *The Guardian* (January 17, 2014), www.theguardian.com/environment/2014/jan/17/1000-rhinos-killed-south-africa-2013.

20. Save the Rhino International, 2014, "South African Poaching Crisis," www.savetherhino.org/support_us/campaigns_and_appeals/586_south_african_poaching_crisis, accessed June 4, 2013.

21. Aislin Laing, 2013, "Last Rhinos in Mozambique Killed by Poachers," *The Telegraph* (April 30, 2013), www.telegraph.co.uk/news/worldnews/africaandindianocean/mozambique/10028738/Last-rhinos-in-Mozambique-killed-by-poachers.html.

22. Paula Kahumbu, 2013, "Seven Rhinos Killed by Poachers in Kenya's Bloodiest Week," *Africa Wild* (June 3, 2013), www.guardian.co.uk/environment/africa-wild/2013/jun/03/rhinos-killed-kenyas-bloodiest-week.

23. IUCN Red List, 2008, *Rhinoceros unicornis,* www.iucnredlist.org/details/19496/0.

24. Hari Kumar, 2013, "Kaziranga Remains a Success Story, Despite Rising Poaching," *New York Times* blog *India Ink* (April 11, 2013), http://india.blogs.nytimes.com/2013/04/11/kaziranga-remains-a-success-story-despite-rising-poaching/. See also "Rhino Falls Prey to Illegal Poaching in Kaziranga National Park," *Zee News* (May 12, 2013), http://zeenews.india.com/news/eco-news/rhino-falls-prey-to-illegal-poaching-in-kaziranga-national-park_848155.html.

25. IUCN Red List, 2008, *Dicerorhinus sumatrensis,* www.iucnredlist.org/details/6553/0.

26. "Indonesia, Malaysia Agree to Save Last 100 Sumatran Rhinos," Environment News Service, April 4, 2013, http://ens-newswire.com/2013/04/04/indonesia-malaysia-agree-to-save-last-100-sumatran-rhinos/, accessed June 9, 2013.

27. IUCN Red List, 2008, *Rhinoceros sondaicus,* www.iucnredlist.org/details/19495/0.

28. Chistopher Lever, 2004, "The Impact of Traditional Chinese Medicine on Threatened Species," *Oryx* 38:3–16.

29. IUCN Red List, 2008, *Ursus thibetanus,* www.iucnredlist.org/details/22824/0.

30. "Bear Bile Extraction," *Journal of Chinese Medicine*, 2013, www.jcm.co.uk/endangered-species-campaign/asiatic-black-bear/bear-bile-extraction/.

31. Andrew Jacobs, 2013, "Folk Remedy Extracted from Captive Bears Stirs Furor in China," *New York Times* (May 21, 2013), www.nytimes.com/2013/05/22/world/asia/chinese-bear-bile-farming-draws-charges-of-cruelty.html.

32. Michael Marshall, 2013, "Elephant Ivory Could Be Bankrolling Terrorist Groups," *New Scientist* (October 2, 2013), www.newscientist.com/article/dn24319-elephant-ivory-could-be-bankrolling-terrorist-groups.html.

33. Elizabeth L. Bennett, 2012, "How to Stop Wildlife Poachers," *New York Times* (May 24, 2012), www.nytimes.com/2012/05/25/opinion/how-to-stop-wildlife-poachers.

34. Jaymi Heimbuch, 2013, "Seven High-Tech Tools to Make Poaching Extinct," *Mother Nature Network* (March 4, 2013), www.mnn.com/earth-matters/animals/stories/7-high-tech-tools-to-make-poaching-extinct.

35. Martin Angler, 2013, "Dye and Poison Stop Rhino Poachers," *Scientific American* blog, May 9, 2013, http://blogs.scientificamerican.com/guest-blog/2013/05/09/dye-and-poison-stop-rhino-poachers/.

36. Duan Biggs, Franck Courchamp, Rowan Martin, and Hugh P. Possingham, 2013, "Legal Trade of Africa's Rhino Horns," *Science* 339:1038–1039; see also their response to letters presenting alternative views, Duan Biggs, Franck Courchamp, Rowan Martin, and Hugh P. Possingham, 2013, "Response," *Science* 340:1168.

37. Allen Collins, Gavin Fraser, and Jen Snowball, 2013, "Rhino Poaching: Supply and Demand Uncertain," *Science* 340:1167; Herbert H.T. Prins and Benson Okita-Ouma, 2013, "Rhino Poaching: Unique Challenges," *Science* 340:1167–1168; Carla A. Litchfield, 2013, "Rhino Poaching: Apply Conservation Psychology," *Science* 340:1168.

38. Andrew Jacobs, 2010, "Tiger Farms in China Feed Thirst for Parts," *New York Times* (February 12, 2010), www.nytimes.com/2010/02/13/world/asia/13tiger.html.

39. NOAA, 2013, "State of the Coast," http://stateofthecoast.noaa.gov/com_fishing/welcome.html.

40. Charles G. Smith and John R. Vane, 2003, "The Discovery of Captopril," *Journal of the Federation of American Societies of Experimental Biology* 17:788–789.

41. R.A. Rodby, L.M. Firth, and E.J. Lewis, 1996, "An Economic Analysis of Captopril in the Treatment of Diabetic Nephropathy: The Collaborative Study Group," *Diabetes Care* 19(10):1051–61.

42. Sustainable Alternatives Network, 2002, Conservation Finance Guide, "Bio-Prospecting: Understanding the Mechanism—How Does It Work?"; published as a joint initiative of the United Nations Environment Programme and the Global Environment Facility, http://conservationfinance.org/guide/guide/indexc2f.htm, accessed June 10, 2013.

43. B. Collen, M. Böhm, R. Kemp, and J.E.M. Baillie, eds., 2012, *Spineless: Status and Trends of the World's Invertebrates* (London: Zoological Society of London); Bradley J. Cardinale, J. Emmett Duffy, Andrew Gonzalez, David U. Hooper, Charles Perrings, Patrick Venail, et al., 2012, "Biodiversity Loss and Its Impact on Humanity," *Nature* 486:59–67.

44. Bradley J. Cardinale, J. Emmett Duffy, Andrew Gonzalez, David U. Hooper, Charles Perrings, Patrick Venail, et al., 2012, "Biodiversity Loss and Its Impact on Humanity," *Nature* 486:59–67.

45. A. Fischlin, 2008, "IPCC Estimates for Emissions from Land-Use Change, Notably Deforestation," Systems Ecology Report No. 31, Terrestrial System Ecology, Institute of Integrative Biology, ETH Zurich, www.sysecol2.ethz.ch/publications/reports/pdfs/fi137.pdf. For emissions and temperature-increase scenarios, see the IPCC AR4 and AR5 reports: IPCC, 2007, "Summary for Policymakers," in *Climate Change 2007: The Physical Science Basis,* ed. S. Solomon, D. Qin, M. Manning, Z. Chen, M. Marquis, K.B. Averyt, M. Tignor, and H.L. Miller (Cambridge: Cambridge University Press), online at www.ipcc.ch/pdf/assessment-report/ar4/wg1/ar4-wg1-spm.pdf; and IPCC, 2013, "Summary for Policymakers," in *Climate Change 2013: The Physical Science Basis,* ed. T.F. Stocker, D. Qin, G.-K. Plattner, M. Tignor, S.K. Allen, J. Boschung, A. Nauels, Y. Xia, V. Bex, and P.M. Midgley (Cambridge: Cambridge University Press), online at www.ipcc.ch/report/ar5/wg1/docs/WGIAR5_SPM_brochure_en.pdf. See note 16 in chapter 3 for a brief explanation of the emissions trajectories specified in each of the IPCC reports.

46. "Carbon Copy," *The Economist* (December 14, 2013), www.economist.com/news/business/21591601-some-firms-are-preparing-carbon-price-would-make-big-difference-carbon-copy.

47. World Land Trust, 2013, "Frequently Asked Questions: Ecosystem Services," www.worldlandtrust.org/eco-services/faq#Q19.

48. Elisabeth Rosenthal, 2009, "New Jungles Prompt a Debate on Rain Forests," *New York Times* (January 29, 2009), www.nytimes.com/2009/01/30/science/earth/30forest.html.

49. Reese Ewing, 2013, "Brazil Soy Farmers, Traders See Smaller Profits as Prices Drop," Reuters (May 6, 2013), www.reuters.com/article/2013/05/06/brazil-basis-soy-idUSL2N0DH2LY20130506.

50. *Soybean and Corn Advisor,* 2009–2011, www.soybeansandcorn.com/Frequently-Asked-Questions, accessed June 11, 2013.

51. Robert Costanza, Ralph d'Arge, Rudolf de Groot, Stephen Farber, Monica Grasso, Bruce Hannon, et al., 1997, "The Value of the World's Ecosystem Services and Natural Capital," *Nature* 387:253–260. The value given for ecosystem services in this paper has been contested, but it certainly gives you a general sense of just how enormous the number is.

52. UN-REDD Programme, 2013, www.un-redd.org/; Environmental Defense Fund, 2013, "Brazil: Conserving Forests to Stop Climate Change," www.edf.org/climate/brazil-conserving-forests-stop-climate-change.

53. "Rainforests as CO2 Storage: Siemens Healthcare Supports WWF Indonesia," *Pharma* blog, October 13, 2010, www.wtgnews.com/2010/10/rain-forests-as-co2-storage-siemens-healthcare-supports-wwf-indonesia-2/.

54. Mark Tercek and Jonathan S. Adams, 2013, *Nature's Fortune: How Business and Society Thrive by Investing in Nature* (New York: Basic Books).

55. Fortune 500 Annual Ranking of America's Largest Corporations, 2013, as reported by CNNMoney, http://money.cnn.com/magazines/fortune/fortune500/2012/snapshots/100.html.

56. Mark Tercek and Jonathan S. Adams, 2013, *Nature's Fortune: How Business and Society Thrive by Investing in Nature* (New York: Basic Books), p. 14.

57. Paul R. Ehrlich, Peter M. Kareiva, and Gretchen C. Daily, 2012, "Securing Natural Capital and Expanding Equity to Rescale Civilization," *Nature* 486:68–73.

58. Gretchen C. Daily, Tore Söderqvist, Sara Aniyar, Kenneth Arrow, Partha Dasgupta, Paul R. Ehrlich, et al., 2000, "The Value of Nature and the Nature of Value," *Science* 289:395–396.

59. Gretchen C. Daily and Katherine Ellison, 2002, *The New Economy of Nature: The Quest to Make Conservation Profitable* (Washington, DC: Island Press); see also Mark Tercek and Jonathan S. Adams, 2013, *Nature's Fortune: How Business and Society Thrive by Investing in Nature* (New York: Basic Books), p. 14.

60. Mark Tercek and Jonathan S. Adams, 2013, *Nature's Fortune: How Business and Society Thrive by Investing in Nature* (New York: Basic Books), p. 14.

61. IUCN Red List of Threatened Species, 2012, *Vultur gryphus*, www.iucnredlist.org/details/106003822/0. Andean condors are classified by the IUCN as near threatened, meaning they are not yet in the vulnerable category but are quickly heading in that direction.

62. Michael Wines, 2013, "Mystery Malady Kills More Bees, Heightening Worry on Farms," *New York Times* (March 28, 2013), www.nytimes.com/2013/03/29/science/earth/soaring-bee-deaths-in-2012-sound-alarm-on-malady.html.

63. Edward Ortiz, 2013, "Honeybee Woes Are Costly for Valley Almond Growers," *Sacramento Bee* (June 13, 2013), www.sacbee.com/2013/05/05/5395636/honeybee-woes-are-costly-for-valley.html; Marc Lifsher, 2012, "Hives for Hire," *Los Angeles Times* (March 3, 2012), articles.latimes.com/2012/mar/03/business/la-fi-california-bees-20120304.

64. Taylor H. Ricketts, Gretchen C. Daily, Paul R. Ehrlich, and Charles D. Michener, 2004, "Economic Value of Tropical Forest to Coffee Production," *Proceedings of the National Academy of Sciences* 101(34):12579–12582.

65. InVEST, Integrated Valuation of Environmental Services and Tradeoffs, the Natural Capital Project, www.naturalcapitalproject.org/InVEST.html; Paul R. Ehrlich, Peter M. Kareiva, and Gretchen C. Daily, 2012, "Securing Natural Capital and Expanding Equity to Rescale Civilization," *Nature* 486:68–73.

66. Joshua H. Goldstein, Giorgio Caldarone, Thomas Kaeo Duarte, Driss Ennaanay, Neil Hannahs, Guillermo Mendoza, Stephen Polasky, Stacie Wolny, and Gretchen C. Daily, 2012, "Integrating Ecosystem-Service Tradeoffs into Land-Use Decisions," *Proceedings of the National Academy of Sciences* 109(19):7565–7570.

67. World Resources Institute, 2013, Mainstreaming Ecosystem Services Initiative (MESI): Tools, www.wri.org/project/mainstreaming-ecosystem-services/tools.

7. RESUSCITATION

1. Mikko Kuussaari, Riccardo Bommarco, Risto K. Heikkinen, Aveliina Helm, Jochen Krauss, Regina Lindborg, et al., 2009, "Extinction Debt: A

Challenge for Biodiversity Conservation," *Trends in Ecology and Evolution* 24(10):564–571; S.T. Jackson and D.F. Sax, 2010, "Balancing Biodiversity in a Changing Environment: Extinction Debt, Immigration Credit and Species Turnover," *Trends in Ecology and Evolution* 25(3):153–160; Wendy B. Foden, Stuart H.M. Butchart, Simon N. Stuart, Jean-Christophe Vié, H. Resit Akçakaya, Ariadne Angulo, et al., 2013, "Identifying the World's Most Climate Change Vulnerable Species: A Systematic Trait-Based Assessment of All Birds, Amphibians and Corals," *PLoS ONE* 8(6):e65427, doi:10.1371/journal.pone.0065427.

2. Aaron Saenz, 2011, "Japanese Scientist to Clone Woolly Mammoth within Five Years!" Singularity HUB, January 11, 2011, http://singularityhub.com/2011/01/19/japanese-scientist-wants-to-clone-a-woolly-mammoth-in-the-next-five-years/.

3. An Asian elephant would be the likely candidate as it belongs to the living species most closely related to woolly mammoths. The two species probably are separated by four million years of evolution, though. For a good entry into the pertinent literature, see Liza Gross, 2006, "Reading the Evolutionary History of the Woolly Mammoth in Its Mitochondrial Genome," *PLoS Biology* 4(3):e74, doi:10.1371/journal.pbio.0040074.

4. O. Handt, M. Hoss, M. Krings, and S. Paabo, 1994, "Ancient DNA: Methodological Challenges," *Experientia* 50(6):524–529; E. Willerslev and A. Cooper, 2005, "Ancient DNA," *Proceedings of the Royal Society* B 272:3–16; T. Lindahl, 2000, "Quick Guide: Fossil DNA," *Current Biology* 10(17):R616; U. Ramakrishnan and E.A. Hadly, 2009, "Using Phylochronology to Reveal Cryptic Population Histories: Review and Synthesis of Four Ancient DNA Studies," *Molecular Ecology* 18:1310–1330.

5. Long Now Foundation, "Revive and Restore Extinct Species Back to Life," http://longnow.org/revive/.

6. Stewart Brand, "The Dawn of De-Extinction: Are You Ready?" TED Talk, February 2013, www.ted.com/talks/stewart_brand_the_dawn_of_de_extinction_are_you_ready.html; "De-Extinction: Bringing Extinct Species Back to Life," *National Geographic*, www.nationalgeographic.com/deextinction/.

7. Stewart Brand, "The Dawn of De-Extinction: Are You Ready?" TED Talk, February 2013, www.ted.com/talks/stewart_brand_the_dawn_of_de_extinction_are_you_ready.html.

8. W.L. Dawson, 1903, *The Birds of Ohio: A Complete Scientific and Popular Description of the 320 Species of Birds Found in the State* (Columbus, OH: Wheaton Publishing), p. 270.

9. Lucy Emerson, 1808, *The New England Cookery, or the Art of Dressing All Kinds of Flesh, Fish, and Vegetables and the Best Modes of Making Puffs, Pies, Tarts, Puddings, Custards, and Preserves, and All Kinds of Cakes, from the Imperial Plumb to Plain Cake, Particularly Adapted to This Part of Our Country* (Montpelier, VT: Printed for Josiah Parks [Proprieter of the Work]).

10. W.L. Dawson, 1903, *The Birds of Ohio: A Complete Scientific and Popular Description of the 320 Species of Birds Found in the State* (Columbus, OH: Wheaton Publishing), p. 425.

11. This account by Professor H.B. Roney was related by W.L. Dawson; ibid., p. 426.

12. Ibid., p. 425.

13. R. Yanagimachi, 2002, "Cloning: Experience from the Mouse and Other Animals," *Molecular and Cellular Endocrinology* 187:241–248, notes that the best success rates are in Japanese cattle, where 20 percent of cloned embryos reach adulthood, but in most animals, 97 percent of cloned embryos die before reaching full term.

14. For more details, see Kelly Servick, 2013, "The Plan to Bring the Iconic Passenger Pigeon Back from Extinction," *Wired* (March 15, 2013), www.wired.com/wiredscience/2013/03/passenger-pigeon-de-extinction/all/.

15. Just assuming a total of $100,000 per year in salaries and benefits hits a million dollars in ten years, and that doesn't count sequencing, gene editing, and other lab costs, which could easily double the salary expense. A yearly salary cost of $100,000 is a low estimate when you consider the portion of a principal investigator's time, plus the cost of even a single grad student, postdoc, or lab tech, whose time would be needed in order to complete the project as quickly as possible. The estimate of around a decade to produce the first passenger pigeon comes largely from information given in the Revive and Restore project's report of discussions at a meeting that participants held at Harvard University. Sequencing the two genomes (band-tailed and passenger pigeon) would, optimistically, take at least a year, more likely a couple. Developing the technology needed to edit the DNA and get it into a germ cell, again thinking optimistically, would be at least two additional years down the road. Editing the band-tailed pigeon DNA to make it mimic passenger pigeon DNA would take another five years or so. Now you need to breed the birds. The idea is that you genetically engineer for the different passenger pigeon traits, one or a couple of traits per generation, checking that each trait you breed for works out as you expect it to. Once you are sure you've got the right genes for the right traits, you begin breeding birds that combine the traits. It's thought it would take at least twelve

gene changes to get 80 percent of the way from a band-tailed pigeon to a passenger pigeon, and maybe thousands to get all the way. Assuming that twelve changes are satisfactory, that egg to next-generation egg takes about six months, and that all twelve strains are bred simultaneously, and allowing for the fact that all seldom goes right the first time around, that's probably another two or three years. Given that sequencing got under way in 2013, it's likely that it will be at least 2023 before we can expect to see what looks like a real passenger pigeon pop out of an egg, assuming everything goes without a hitch.

16. See Kelly Servick, 2013, "The Plan to Bring the Iconic Passenger Pigeon Back from Extinction," *Wired* (March 15, 2013), www.wired.com/wiredscience/2013/03/passenger-pigeon-de-extinction/all/.

17. European starlings were introduced into New York's Central Park in 1890 and 1891 and over the next sixty years spread rapidly all the way to the Pacific Coast.

18. Carl Zimmer, 2013, "Bringing Them Back to Life," *National Geographic* (July 2013), http://ngm.nationalgeographic.com/2013/04/125-species-revival/zimmer-text.

19. D. Nogués-Bravo, J. Rodríguez, J. Hortal, P. Batra, and M.B. Araújo, 2008, "Climate Change, Humans, and the Extinction of the Woolly Mammoth," *PLoS Biology* 6(4):e79, doi:10.1371/journal.pbio.0060079.

20. P.L. Koch and A.D. Barnosky, 2006, "Late Quaternary Extinctions: State of the Debate," *Annual Review of Ecology, Evolution, and Systematics* 37:215–250.

21. Stuart Pimm, 2013, "The Case against Species Revival," *National Geographic* (March 12, 2013), http://news.nationalgeographic.com/news/2013/03/130312-deextinction-conservation-animals-science-extinction-biodiversity-habitat-environment/. See also Corey Bradshaw, 2013, "De-Extinction Is About as Sensible as De-Death," *The Conversation* (March 15, 2013), http://theconversation.com/de-extinction-is-About-as-sensible-as-de-death-12850; and Liza Gross, 2013, "De-Extinction Debate: Should Extinct Species Be Revived?" *KQED Science* (June 5, 2013), http://blogs.kqed.org/science/2013/06/05/deextinction-debate-should-extinct-species-be-revived/.

22. E.A. Hadly & A.D. Barnosky, 2009, "Vertebrate Fossils and the Future of Conservation Biology," *Conservation Paleobiology: Using the Past to Manage for the Future,* Paleontological Society Papers 15:39–59.

23. A.S. Leopold, S.A. Cain, C.M. Cottam, I.N. Gabrielson, and T. Kimball, 1963, "Wildlife Management in the National Parks: The Leopold Report," www.nps.gov/history/history/online_books/leopold/leopold.htm.

24. Society of Conservation Biology Bylaws, www.conbio.org/about-scb/who-we-are/bylaws-of-scb#A4; C.M. Meine, M. Soulé, and R.F. Noss, 2006, "'A Mission-Driven Discipline': The Growth of Conservation Biology," *Conservation Biology* 20(3):631–651.

25. C.M. Meine, M. Soulé, and R.F. Noss, 2006, "'A Mission-Driven Discipline': The Growth of Conservation Biology," *Conservation Biology* 20(3):631–651.

26. Michael Hoffmann, Craig Hilton-Taylor, Ariadne Angulo, Monika Böhm, Thomas M. Brooks, Stuart H.M. Butchart, et al., 2010, "The Impact of Conservation on the Status of the World's Vertebrates," *Science* 330:1503–1509.

27. Ken Cole, Lara Schmit, and Catherine Puckett, 2011, "Uncertain Future for Joshua Trees Projected with Climate Change," United States Geological Survey News Room, March 24, 2011, www.usgs.gov/newsroom/article.asp?ID = 2723; Kenneth L. Cole, Kirsten Ironside, Jon Eischeid, Gregg Garfin, Phillip B. Duffy, and Chris Toney, 2011, "Past and Ongoing Shifts in Joshua Tree Distribution Support Future Modeled Range Contraction," *Ecological Applications* 21:137–149.

28. Michelle Marvier, Peter Kareiva, and and Robert Lalasz, 2012, "Conservation in the Anthropocene," *Breakthrough Journal,* www.rightsandresources.org/documents/files/doc_5337.pdf; see also Peter Kareiva and Michelle Marvier, 2011, *Conservation Science: Balancing the Needs of People and Nature* (Greenwood Village, CO: Roberts).

29. Emma Marris, 2012, *Rambunctious Garden: Saving Nature in a Post-Wild World* (New York: Bloomsbury).

30. For arguments against many of the points made in articles about the "new" conservation biology, see, for example, Kieran Suckling, 2012, "Conservation for the Real World," *Breakthrough Journal,* http://thebreakthrough.org/index.php/journal/debates/conservation-in-the-anthropocene-a-breakthrough-debate/conservation-for-the-real-world/; and Andrew C. Revkin, 2012, "Critic of Conservation Efforts Gets Critiqued," *New York Times* blog *Dot Earth,* http://dotearth.blogs.nytimes.com/2012/04/10/peter-kareiva-critic-of-environmentalism-gets-critiqued/.

31. Michelle Marvier, Peter Kareiva, and Robert Lalasz, 2012, "Conservation in the Anthropocene," *Breakthrough Journal,* www.rightsandresources.org/documents/files/doc_5337.pdf.

32. James William Gibson, 2013, "Is the Far Right Driving Gray Wolves to Extinction?" *Salon* (June 25, 2013), www.salon.com/2013/06/25/is_the_far_right_driving_gray_wolves_to_extinction_partner/singleton/; Dan Frosch, 2013, "Report Criticizes U.S. Stewardship of Wild Horses," *New York Times* (June 6,

2013), www.nytimes.com/2013/06/07/us/report-criticizes-us-stewardship-of-wild-horses.html.

33. The value of places that are little influenced by humans has been widely written about by scientists as well as people who deal mainly with emotions. See, for example, writings by Aldo Leopold and Michael Soulé for scientists' views on nature, and those by Wallace Stegner and Terry Tempest Williams for views of humanists. People have defined nature in many, many different ways—there are almost as many definitions as people who write about it—but the key point is that a broad swath of society values, as an essential kind of nature, ecosystems that are as little influenced by humans as is possible on a planet with seven billion people.

34. Emma Marris makes this same point in the closing chapter of her 2012 book *Rambunctious Garden: Saving Nature in a Post-Wild World* (New York: Bloomsbury).

35. J.S. McLachlan and J.J. Hellmann, 2007, "A Framework for the Debate of Assisted Migration in an Era of Climate Change," *Conservation Biology* 21:297–302; R. Early and D. Sax, 2011, "Analysis of Climate Paths Reveals Potential Limitations on Species Range Shifts," *Ecology Letters* 14:1125–1133.

36. Anthony D. Barnosky, 2009, *Heatstroke: Nature in an Age of Global Warming* (Washington, DC: Island Press).

37. C.J. Donlan, 2005, "Re-Wilding North America," *Nature* 436(7053):913–914; C.J. Donlan, J. Berger, C.E. Bock, J.H. Bock, D.A. Burney, J.A. Estes, et al., 2006, "Pleistocene Rewilding: An Optimistic Agenda for Twenty-First Century Conservation," *American Naturalist* 168(5):660–681; Paul S. Martin, 2005, *Twilight of the Mammoths: Ice Age Extinctions and the Rewilding of America* (Berkeley: University of California Press).

38. James Gorman, 2013, "To Save the King of the Jungle, a Call to Pen Him In," *New York Times* (March 26, 2013), www.nytimes.com/2013/03/27/science/lion-researchers-call-for-more-fences-to-save-the-big-cats.html.

8. BACK FROM THE BRINK

1. Anthony D. Barnosky, James H. Brown, Gretchen C. Daily, Rodolfo Dirzo, Anne H. Ehrlich, Paul R. Ehrlich, et al., 2014, "Introducing the *Scientific Consensus on Maintaining Humanity's Life Support Systems in the 21st Century: Information for Policy Makers*," *Anthropocene Review* 1:78–109, doi:10.1177/2053019613516290, with background information and foreign-language versions available at http://consensusforaction.stanford.edu/.

2. *The Guardian,* June 12, 2013, http://m.guardiannews.com/environment/2013/jun/12/european-coal-pollution-premature-deaths.

3. DownToEarth, June 23, 2013, www.downtoearth.org.in/content/corpses-rot-flow-down-ganga-uttarakhand.

4. NBC News, June 13, 2013, http://usnews.nbcnews.com/_news/2013/06/13/18932265-two-killed-as-colorado-wildfires-destroy-360-homes-force-evacuations-in-colorado-springs.

5. CNN News, June 18, 2013, www.cnn.com/2013/06/18/us/texas-drought-emergency.

6. This statistic is for the year 2010, as reported in Stephen S. Lim, Theo Vos, Abraham Flaxman, Goodarz Danaei, Kenji Shibuya, Heather Adair-Rohani, et al., 2012, "A Comparative Risk Assessment of Burden of Disease and Injury Attributable to 67 Risk Factors and Risk Factor Clusters in 21 Regions, 1990–2010: A Systematic Analysis for the Global Burden of Disease Study 2010," *Lancet* 380:2224–2260.

7. See www1.ncdc.noaa.gov/pub/data/papers/smith-and-katz-2013.pdf.

8. Adam B. Smith and Richard B. Katz, 2013, "U.S. Billion-Dollar Weather and Climate Disasters: Data Sources, Trends, Accuracy and Biases," *Natural Hazards* 67:387–410.

9. NOAA National Climatic Data Center, 2013, "NCDC Releases 2012 Billion-Dollar Weather and Climate Disasters Information," www.ncdc.noaa.gov/news/ncdc-releases-2012-billion-dollar-weather-and-climate-disasters-information.

10. Eric Lambin, 2012, *An Ecology of Happiness* (Chicago: University of Chicago Press).

11. See p. 11 in IPCC, 2012, "Summary for Policymakers," in *Managing the Risks of Extreme Events and Disasters to Advance Climate Change Adaptation,* ed. C. B. Field, V. Barros, T. F. Stocker, D. Qin, D. J. Dokken, K. L. Ebi, et al. (Cambridge: Cambridge University Press), pp. 1–19, online at http://ipcc-wg2.gov/SREX/images/uploads/SREX-SPMbrochure_FINAL.pdf.

12. S. C. Sherwood and M. Huber, 2010, "An Adaptability Limit to Climate Change Due to Heat Stress," *Proceedings of the National Academy of Sciences* 107:9552–9555.

13. See p. 11 in IPCC, 2012, "Summary for Policymakers," in *Managing the Risks of Extreme Events and Disasters to Advance Climate Change Adaptation,* ed. C. B. Field, V. Barros, T. F. Stocker, D. Qin, D. J. Dokken, K. L. Ebi, et al. (Cambridge: Cambridge University Press), pp. 1–19, online at http://ipcc-wg2.gov/SREX/images/uploads/SREX-SPMbrochure_FINAL.pdf.

14. Thomson Reuters Foundation, 2013, "Pakistan Floods," April 8, 2013, www.trust.org/spotlight/Pakistan-floods-2010.

15. Robert I. McDonald, Pamela Green, Deborah Balk, Balazs M. Fekete, Carmen Revenga, Megan Todd, and Mark Montgomery, 2011, "Urban Growth, Climate Change, and Freshwater Availability," *Proceedings of the National Academy of Sciences* 108:6312–6317.

16. AllAfrica, June 18, 2013, http://allafrica.com/stories/201306181489.html.

17. *Belfast Telegraph,* June 4, 2013, www.belfasttelegraph.co.uk/news/local-national/uk/uk-a-few-days-away-from-food-shortage-29319442.html.

18. *The Guardian,* October 13, 2012, www.guardian.co.uk/global-development/2012/oct/14/un-global-food-crisis-warning.

19. Science 2.0, June 20, 2013, www.science20.com/news_articles/legacy_food_production_techniques_wont_feed_population_2050–115171, reporting on the study by D.K. Ray, N.D. Mueller, P.C. West, and J.A. Foley, 2013, "Yield Trends Are Insufficient to Double Global Crop Production by 2050," *PLoS ONE* 8(6):e66428, doi:10.1371/journal.pone.0066428.

20. World Health Organization, Millennium Development Goals, "MDG 1: Eradicate Extreme Poverty and Hunger," www.who.int/topics/millennium_development_goals/hunger/en/index.html.

21. Food and Agriculture Organization of the United Nations, 2012, *The State of Food Insecurity in the World 2012*, www.fao.org/publications/sofi/en/.

22. *The Guardian,* October 13, 2012, www.guardian.co.uk/global-development/2012/oct/14/un-global-food-crisis-warning.

23. D.K. Ray, N.D. Mueller, P.C. West, and J.A. Foley, 2013, "Yield Trends Are Insufficient to Double Global Crop Production by 2050," *PLoS ONE* 8(6):e66428, doi:10.1371/journal.pone.0066428.

24. Sharon M. Gourdji, Adam M. Sibley, and David B. Lobell, 2013, "Global Crop Exposure to Critical High Temperatures in the Reproductive Period: Historical Trends and Future Projections," *Environmental Research Letters* 8:8 024041, doi:10.1088/1748–9326/8/2/024041.

25. A. Shah, 2013, "Poverty Facts and Stats," *Global Issues,* www.globalissues.org/article/26/poverty-facts-and-stats.

26. USDA Economic Research Service, www.ers.usda.gov/topics/food-nutrition-assistance/food-security-in-the-us/key-statistics-graphics.aspx#insecure. The number of people in the United States living below poverty level was reported by CBS to be 49.7 million in 2011: "U.S. Poverty Rate Spikes, Nearly 50 Million Americans Affected," November 15, 2012, http://washington.cbslocal.com/2012/11/15/census-u-s-poverty-rate-spikes-nearly-50-million-

americans-affected/. The number reported by the U.S. Census Bureau and other news outlets was 46.2 million; see, for example, Heidi Shierholz and Elise Gould, 2012, "Already More Than a Lost Decade: Poverty and Income Trends Continue to Paint a Bleak Picture," Economic Policy Institute, September 12, 2012, www.epi.org/publication/lost-decade-poverty-income-trends-continue-2/. Depending on the size of a family, the U.S. poverty threshold ranges between $13 and $30 per person per day, as specified by the U.S. Department of Health and Human Services 2011 guidelines, http://aspe.hhs.gov/poverty/11poverty.shtml.

27. Laurence Chandy, Natasha Ledlie, and Veronika Penciakov, 2013, "The Final Countdown: Prospects for Ending Extreme Poverty by 2030," Brookings Institution Policy Paper 2013–04, www.brookings.edu/~/media/Research/Files/Reports/2013/04/ending%20extreme%20poverty%20chandy/The_Final_Countdown.pdf.

28. Country Income Groups, World Bank Classification, http://chartsbin.com/view/2438; World Bank, 2013, "How We Classify Countries," http://data.worldbank.org/about/country-classifications.

29. Laurence Chandy and Geoffrey Gertz, 2011, "Two Trends in Global Poverty," Brookings Institution, May 17, 2011, www.brookings.edu/research/opinions/2011/05/17-global-poverty-trends-chandy.

30. Ibid.

31. GSDRC Applied Knowledge Systems, 2013, *Fragile States,* chapter 1, "Understanding Fragile States," www.gsdrc.org/go/fragile-states/chapter-1--understanding-fragile-states.

32. Organisation for Economic Co-operation and Development, Development Assistance Committee, 2012, "Fragile States 2103: Resource Flows and Trends in a Shifting World," www.oecd.org/dac/incaf/FragileStates2013.pdf.

33. The Organisation for Economic Co-operation and Development (OECD) includes the United States, Canada, Mexico, Australia, Japan, most western European countries, some from eastern Europe, and Iceland; for the complete list, see www.oecd.org/about/membersandpartners/. David Lynch, 2011, "Growing Income Gap May Leave U.S. Vulnerable," Bloomberg, October 13, 2011,www.bloomberg.com/news/2011–10–13/growing-income-divide-may-increase-u-s-vulnerability-to-financial-crises.html.

34. OECD, 2011, "Income Inequality," in *OECD Factbook 2011–2012: Economic, Environmental and Social Statistics,* doi:10.1787/factbook-2011–31-en, summarized at www.oecd.org/els/soc/income-distribution-database.htm, and "An Overview of Growing Income Inequalities in OECD Countries: Main Findings, www.oecd.org/els/soc/49499779.pdf.

35. Linette Lopez, 2011, "This Is How Income Inequality Destroys Societies," *Business Insider* (November 1, 2011), www.businessinsider.com/the-negative-effects-of-income-inequality-on-society-2011–11.

36. Dion Harmon, Blake Stacey, Yavni Bar-Yam, and Yaneer Bar-Yam, 2010, "Networks of Economic Market Interdependence and Systemic Risk," Cornell University Library arXiv:1011.3707v2, http://arxiv.org/pdf/1011.3707v2.pdf.

37. Anthony D. Barnosky, James H. Brown, Gretchen C. Daily, Rodolfo Dirzo, Anne H. Ehrlich, Paul R. Ehrlich, et al., 2014, "Introducing the *Scientific Consensus on Maintaining Humanity's Life Support Systems in the 21st Century: Information for Policy Makers*," *Anthropocene Review* 1:78–109, doi:10.1177/2053019613516290, with background information and foreign-language versions available at http://consensusforaction.stanford.edu/.

38. For example, within seven years during World War II, the United States scaled up from 3,000 to 300,000 airplanes, and within fifty years it built enough roads to encircle Earth twice.

39. Anthony D. Barnosky, James H. Brown, Gretchen C. Daily, Rodolfo Dirzo, Anne H. Ehrlich, Paul R. Ehrlich, et al., 2014, "Introducing the *Scientific Consensus on Maintaining Humanity's Life Support Systems in the 21st Century: Information for Policy Makers*," *Anthropocene Review* 1:78–109, doi:10.1177/2053019613516290, with background information and foreign-language versions available at http://consensusforaction.stanford.edu/.

40. Naomi Oreskes and Erik M. Conway, 2010, *Merchants of Doubt* (New York: Bloomsbury).

41. For 2008, see http://money.cnn.com/magazines/fortune/fortune500/2008/performers/companies/profits/; 2009, http://money.cnn.com/magazines/fortune/fortune500/2009/performers/companies/profits/; 2010, http://money.cnn.com/magazines/fortune/fortune500/2010/performers/companies/profits/; 2011, http://money.cnn.com/magazines/fortune/fortune500/2011/full_list/; 2012, http://money.cnn.com/magazines/fortune/fortune500/2012/performers/companies/profits/; 2013, http://money.cnn.com/gallery/magazines/fortune/2013/07/08/global-500-most-profitable.fortune/.

42. Ibid.

43. *The Day after Tomorrow* was a 2004 movie that depicted the results of global warming, including rising sea levels and intense storms that decimated New York. Its major premise, that global warming would cause changes in ocean currents that turned New York frigid, was not supported by science, but the devastation of New York City stuck in the public imagination.

44. See, for example, A. Leiserowitz, N. Smith, and J.R. Marlon, 2010, "Americans' Knowledge of Climate Change," Yale University, Yale Project on Climate Change Communication, http://environment.yale.edu/climate/files/ClimateChangeKnowledge2010.pdf; and David Perlman, 2010, "Stanford Survey Finds More Doubt Global Warming," *San Francisco Chronicle* (June 10, 2010), www.sfgate.com/science/article/Stanford-survey-finds-more-doubt-global-warming-3261848.php.

45. See http://climate.nasa.gov/scientific-consensus and W.R.L. Anderegg, 2010, "Expert Credibility in Climate Change," *Proceedings of the National Academy of Sciences* 1(1):12107–12109, doi:10.1073/pnas.1003187107; P.T. Doran and M.K. Zimmerman, 2009, "Examining the Scientific Consensus on Climate Change," *Eos Transactions American Geophysical Union* 90(3):22, doi:10.1029/2009EO030002; N. Oreskes, "Beyond the Ivory Tower: The Scientific Consensus on Climate Change," *Science* 306:1686, doi:10.1126/science.1103618.

46. *The Globe and Mail,* June 8, 2013, "Obama, Xi Agree to Fight Climate Change," www.theglobeandmail.com/news/world/obama-xi-agree-to-fight-climate-change/article12439864/.

47. Justin Gillis, 2009, "Norman Borlaug, Plant Scientist Who Fought Famine, Dies at 95," *New York Times* (September 13, 2009), www.nytimes.com/2009/09/14/business/energy-environment/14borlaug.html.

48. Henry I. Miller, 2012, "Norman Borlaug: The Genius behind the Green Revolution," *Forbes* (January 18, 2012), www.forbes.com/sites/henrymiller/2012/01/18/norman-borlaug-the-genius-behind-the-green-revolution/.

49. AgBioWorld, 2011, "Political Aspects of the Green Revolution," www.agbioworld.org/biotech-info/topics/borlaug/political.html.

50. Henry I. Miller, 2012, "Norman Borlaug: The Genius behind the Green Revolution," *Forbes* (January 18, 2012), www.forbes.com/sites/henrymiller/2012/01/18/norman-borlaug-the-genius-behind-the-green-revolution/.

51. Amanda Briney, 2008, "Green Revolution: History and Overview of the Green Revolution," About.com Geography, October 23, 2008, http://geography.about.com/od/globalproblemsandissues/a/greenrevolution.htm.

52. AgBioWorld, 2011, "Political Aspects of the Green Revolution," www.agbioworld.org/biotech-info/topics/borlaug/political.html.

53. Ibid.

54. Henry I. Miller, 2012, "Norman Borlaug: The Genius behind the Green Revolution," *Forbes* (January 18, 2012), www.forbes.com/sites/henrymiller/2012/01/18/norman-borlaug-the-genius-behind-the-green-revolution/.

55. Tom Hargrove and W. Ronnie Coffman, 2006, "Breeding History," *Rice Today* October-December 2006:35–38, www.goldenrice.org/PDFs/Breeding_History_Sept_2006.pdf.

56. World Business Council for Sustainable Development, 2013, "Business Solutions for a Sustainable World," www.wbcsd.org/about/members.aspx.

57. World Business Council for Sustainable Development, 2013, "Vision 2050," www.wbcsd.org/vision2050.aspx, and "From Vision 2050 to Action 2020," www.wbcsd.org/action2020.aspx.

58. John Funk, 2013, "Exxon's Rex Tillerson Sees Climate Change as Risk Management Problem," *The Plain Dealer* (June 14, 2013), www.cleveland.com/business/index.ssf/2013/06/exxons_rex_tillerson_see_clima.html.

59. www.montereybayaquarium.org/cr/cr_seafoodwatch/sfw_recommendations.aspx.

60. A good place to start figuring out where you want to spend your money is to look at the World Business Council for Sustainable Development website, www.wbcsd.org/home.aspx. There you will find a list of participating corporations, and from that, you can burrow deeper into the internet to figure out which are modifying their business plans toward environmental sustainability. A good short list is the companies actively participating in the Vision 2050 and Action 2020 plans.

61. Appendices I, II, and III, Convention on International Trade in Endangered Species of Wild Fauna and Flora, www.cites.org/eng/app/appendices.php.

62. California Office of the Governor Newsroom, 2013, "Governor Brown Expands Partnership with China to Combat Climate Change," http://gov.ca.gov/news.php?id=18205.

63. California Office of the Governor Newsroom, 2013, "Governor Brown Joins Oregon, Washington, British Columbia Leaders to Combat Climate Change," http://gov.ca.gov/news.php?id=18284.

64. "Fact Sheet: Executive Order on Climate Preparedness," November 1, 2013,www.whitehouse.gov/the-press-office/2013/11/01/fact-sheet-executive-order-climate-preparedness.

65. Stuart L. Pimm, 2002, *The World According to Pimm: A Scientist Audits the Earth* (Rutgers, NJ: Rutgers University Press).

66. P.R. Ehrlich, P. Kareiva, and G.C. Daily, 2012, "Securing Natural Capital and Expanding Equity to Rescale Civilization," *Nature* 486:68–73; J.J. Speidel, D.C. Weiss, S.A. Ethelston, and S.M. Gilbert, 2009, "Population Policies,

Programmes and the Environment," *Philosophical Transactions of the Royal Society* A 364:3049–3065, doi:10.1098/rstb.2009.0162; Mary K. Shenk, Mary C. Towner, Howard C. Kress, and Nurul Alam, 2013, "A Model Comparison Approach Shows Stronger Support for Economic Models of Fertility Decline," *Proceedings of the National Academy of Sciences* 110(20):8045–8050, www.pnas.org/content/early/2013/04/25/1217029110.full.pdf+html.

67. See, for example, the growing company d.light, which has brought off-grid lights to more than 24,000,000 people: www.dlightdesign.com/.

Index

Abbey, Edward, 147
ACE inhibitors, 121–22
acidification of oceans, 37, 46–47, 48
action. *See* solutions
Action 2020, 168–69
Adams, Jonathan, 126
Africa: bush-meat trade, 93–94; elephants in, 10–11, 106–8; food insecurity in, 93, 155; rhinoceros in, 112–14; Sahel desertification, 85. *See also specific African countries and animals*
Age of Dinosaurs, 21
agriculture, 36, 80–93; aquaculture, 98–101, 103; bee declines and, 129; biofuels production, 69, 70–71, 92, 193n35, 200–201n35; Brazilian soybean moratorium, 126; climate disruption and, 90, 156; effective, sustainable agriculture solutions, 88–93, 164–68; extent of current land use, 80–81, 196n3, 200–201n35; food vs. fuel production, 70, 193n35; forest conversions, 84, 88–89, 122–24, 170, 198–99n22; fossil fuels inputs, 193n35; Green Revolution, 89, 164–68; for hydrogen fuel production, 72; impacts on non-human species, 81–88, 112–13; per capita land use figures, 88–89; pesticide use, 89, 90, 166, 167; pollution from, 48, 91, 94, 99; reducing carbon emissions from, 63, 64; water use and conservation, 90–91; wildlife farming, 93, 119–20, 149; yields and yield gap, 89, 90, 155, 156, 164–68. *See also* fertilizer use; food; livestock; pasturelands
ahi (yellowfin tuna), 97–98, 105
air pollution, 152, 153. *See also* emissions
Alamosaurus, 21
albacore tuna, 97, 103
Albertosaurus, 21
algae, 46, 48, 52; algae-based biofuels, 70–72, 92
ALL Species Foundation, 135
alternative energy sources. *See* carbon-neutral energy technologies; *specific types*
Alvarez, Luis, 22. *See also* asteroid-impact extinction theories
Alvarez, Walter, 21–22, 24, 26. *See also* asteroid-impact extinction theories
amphibians, 11, 30–31, 32–33, 144

Amur tiger, 110, 111
Andean condor, 128, 211n61
angiosperms, 39
anoxia, 55; dead zones, 47–48, 71, 94, 99
antelope, 116
apatite, 45
aquaculture, 98–101, 103
aquatic ecosystems. *See* fish and fisheries; lakes; marine *entries*; oceans
AR_4 report (IPCC), 185n16
AR_5 report (IPCC), 185n16
Archibald, J. David, 25
Asaro, Frank, 22. *See also* asteroid-impact extinction theories
Asian elephant, 134, 141, 212n3
Asian markets, for illicit wildlife trade, 109, 110, 113
Asian wildlife: poaching and trade in, 109–11, 115, 116–17. *See also specific animals*
Asiatic black bear, 116–17, 120
asteroid-impact extinction theories, 21–28, 38, 181–82n7
Atlantic bluefin tuna, 11, 96, 97, 102
Atlantic cod, 47
Atlantic salmon, 98–100, 103
atmospheric CO_2. *See* CO_2 levels
aviation, biofuels for, 69–70, 71–72

bacteria, 52
Balaenoptera musculus (blue whale), 94
Bali tiger, 110
band-tailed pigeon, 138, 139, 213–14n15
battery technology, 65, 73, 76, 77
bears: Asiatic black bear, 116–17, 120; grizzly, 145, 148
bees, 129, 130
Bengal tiger, 110
bigeye tuna, 97
Big Five Mass Extinctions, 13, 29–33, 48. *See also* K-Pg event; P-T event
binomial nomenclature, 5–6, 7
biodiversity: of agricultural and pasturelands, 81, 84–88; crop diversity, 90, 167; ecosystem services and, 122; predators and, 82, 197n7
biodiversity education, 168
biofuels, 64, 69–72, 74–75, 92, 193n35
biomass, energy distribution and, 56–58
biotechnology, 70, 90. *See also* de-extinction; molecular biology
birds, 4, 26, 30–31, 32–33, 182–83n17. *See also individual species*
The Birds of Ohio (Dawson), 136, 137
bizarre-nosed chameleon, 11, 12
blackfin tuna, 98
black-market wildlife trade. *See* wildlife poaching and trade
black rhinoceros, 112–13
blood pressure medications, 120–22
blue-capped hummingbird, 11
bluefin tuna, 11, 96–97, 102, 103, 105
blue whale, 94
Borlaug, Norman, 165–67. *See also* Green Revolution
brachiopods, 40, 46
Brand, Stewart, 135–36; Revive and Restore project, 135–41
Brazil: fer-de-lance viper and ACE inhibitor development, 120–22; rainforest conversion economics, 122–24; REDD carbon trading program, 125
British Columbia climate-change initiatives, 172
Bromus tectorum (cheatgrass), 85–88
business, 126, 131, 171–72; climate-change denial and, 161–62; integrating ecosystem services valuation into, 124–25, 131, 146, 149, 168, 169; Vision 2050, 168–69, 222n60. *See also* consumer awareness and choice

$\delta^{13}C_{carb}$, 41, 42–43, 184n8
California: Chinook salmon aquaculture, 100; government climate-change initiatives, 172

Calumma hafahafa (bizarre-nosed chameleon), 11, 12
Cameroon, elephant poaching in, 10
Canadian lynx, 96, 202–3n51
captive breeding programs, 15
Captopril, 121–22
carbon capture and storage (CCS), 63, 64, 74–75
carbon dioxide. *See* CO_2 levels; emissions *entries*; energy *entries*
carbon-neutral energy technologies, 62, 63, 64–66, 75–78, 172; biofuels, 64, 69–72; scaling up, 75–77. *See also specific types*
carbon sequestration, by forests, 123
carbon trading, 124–25
Cargill, 126
Cascade Investment, 71
Caspian tiger, 110
CCS (carbon capture and storage), 63, 64, 74–75
Ceratotherium simum (white rhinoceros), 112, 113–14, 147
certification programs, 100, 101
cetaceans, 94, 202n46
chameleon, bizarre-nosed, 11, 12
change: rate of ecological change, 142, 143–45, 149; resistance to, 77, 124, 161, 165–66. *See also* solutions
Charles Darwin Foundation, 14–15, 16
cheatgrass, 85–88
cheetahs, 150
Chelonoidis nigra (Galápagos tortoise), *xiv*, 1–4, 8, 10, 14–16
Chicxulub Crater, 25–26. *See also* asteroid-impact extinction theories; K-Pg event
chimpanzees, 93
China: deforestation and reforestation in, 127; as market for illicit wildlife trade, 113, 116–17; recent climate-change initiatives, 163, 169; wildlife farming in, 120
Chinese medicine, illicit wildlife trade and, 109–17; Asiatic black bear, 116–17, 120; rhinoceros, 109, 111–15, 118–19; tigers, 109–11, 116
Chinook salmon, 100
Chu, Steven, 65–66, 191n21
cichlids, 100, 101
CITES (Convention on International Trade in Endangered Species of Wild Fauna and Flora), 15, 171, 180n15
Clemens, William A., 24–25, 26
climate change and disruption, 60, 144, 147; agriculture and, 90, 156; conservation approaches and, 145, 146; current impacts on humans, 34–35, 85, 152–55; denial and public opinion about, 161–65; desertification and, 85; economic impacts, 34, 153; extreme weather, 34–35, 36, 37, 153–55, 156, 163; as factor in past extinctions, 28, 41–47, 48, 141; impacts on nonhuman species and habitats, 60, 63, 78, 85, 144, 190n18; recent governmental initiatives, 163, 169, 172; sea level rises, 37, 154. *See also* warming
climate regulation, 122, 123
CO_2 emissions. *See* emissions; energy; vehicle emissions
CO_2 levels: current levels, 185–86n17; IPCC projections, 186n24; ocean acidification and, 37; past, measuring/estimating, 40–42, 184n8; Permian rise and its impacts, 38, 40–47, 184–85n15
coal, 54; coal-fired power plants, 64, 73–75, 153; coal mining, 50–52, 58–59, 78; emissions from, 61, 73–74, 152, 153, 188n8, 194n47. *See also* fossil fuels
Coca-Cola, 126
cod, 103
coffee farms, 129–30
colony collapse disorder, 129
Colorado: author's early life in, 50–51, 78, 143–44; forests, 78, 144

Columba livia (rock pigeon), 139–40
communications technologies, 49, 116, 117, 160–61
condor, Andean, 128, 211n61
conodonts, 44–45
conservation biology, 133; de-extinction opposition, 142; management focus, 147–48, 149–50, 151; new directions and perspectives in, 132–33, 146–48, 149–51; "protect and preserve" ethic, 143–46, 149; rewilding, 150. *See also* de-extinction; wildlife conservation
conservation tillage, 64
consumer awareness and choice, 126, 131, 171, 222n60; food choices, 89, 91–92, 100, 101, 103–5, 170; illicit wildlife trade and, 116–17, 118–20
Convention on International Trade in Endangered Species of Wild Fauna and Flora (CITES), 15, 171, 180n15
Conway, Erik, 161
corals, 40, 46
Corbett's tiger, 110
Costa Rica: coffee farms in, 129–30; government ecosystem services payment initiatives, 127
coyotes, 82
Cretaceous, 20. *See also* K-Pg event
"Cretaceous Barbeque" paper, 27
crinoids, 40
crocodiles, 26
crop diversity, 90, 167
croplands. *See* agriculture
cyanobacteria, 72
cycads, 4

Daily, Gretchen, 129–30
damselfish, 47
Darwin, Charles, 3–4
Darwin Foundation, 14–15, 16
Davis, Steven J., 65
Dawson, W. L., 136, 137
The Day after Tomorrow, 163, 220n43
dead zones, 47–48, 71, 94, 99
decomposition, 48, 184–85n15
deer, 116
de-extinction, 133–42, 149; costs and time required, 140, 213–14n15; feasibility challenges, 134–35, 138–39, 213–14n15; mammoth cloning, 133–35, 141, 212n3; Revive and Restore passenger pigeon project, 136–41, 213–14n15; shortcomings and opposition, 142
deforestation: conversions to agriculture, 84, 88–89, 122–24, 170, 193n35, 198–99n22; costs of, 127; economics and disincentives, 122–24, 125, 127; Malayan tapir and, 84; rainforests, 88–89, 170, 198–99n22; reducing/reversing, 64, 125, 127, 168
Delucchi, Mark, 75–76
Democratic Republic of the Congo, 157; wildlife and poaching in, 10, 106, 113, 147
desertification, 85
developing countries: carbon trading in, 125; poverty in, 157; residents as part of solution, 174–75
Devonian: Late Devonian extinction, 31, 48
diapsids, 39, 183n5
Dicerorhinus sumatrensis (Sumatran rhinoceros), 115
Diceros bicornis (black rhinoceros), 112–13
Dicynodon, 39
dinosaurs, 13, 21. *See also* K-Pg event
DNA: DNA-based species identification, 104–5, 119; extinct species cloning efforts, 133–34, 138–39
DNA Direct, 135
dolphins, 202n46
drought, 37, 153, 154, 156

echinoderms, 46
ecological change: rate of, 142, 143–45, 149. *See also* climate change

economic inequality, 156–59
economic policy, 131; agricultural policy, 166–67; bush-meat hunting disincentives, 93; deforestation disincentives, 124–25; emissions reduction incentives, 77, 78; for water conservation, 91. *See also* public policy initiatives
economic solutions, 168–69; integrating ecosystem services valuation, 124–31, 146, 149, 168, 169. *See also* consumer awareness; money
economic systems: economic instability, 158–59; food trade and food reserves, 155–56; growth and impacts of inequality, 157–59; integrating ecosystem services valuation, 124–31, 146, 149, 168, 169; short-term vs. long-term exploitation of natural capital, xi, 108–9, 120, 122–24. *See also* business
ecosystems conservation, 142; "protect and preserve" ethic, 143–46, 149. *See also* conservation biology
ecosystem services, 106, 109; integrating value into existing economic system, 124–31, 146, 149, 168, 169; valuation of, 108–9, 120–24, 129–31; valuation tools, 130–31. *See also* natural capital
ecotourism, 107–9, 147
Ectopistes migratorius (passenger pigeon), 136–41, 213–14n15
Ecuador: Quito Water Fund, 128–29. *See also* Galápagos
education, 168
education access, 175
efficiency improvements: energy use, 63, 64, 66–69, 73, 168; food production, 88–93, 164–68, 200–201n35
Ehrlich, Paul, 129–30
electricity generation and delivery, 75, 76–77, 172. *See also* energy production; power plants
electricity use, 64
electric vehicles, 64, 65, 72–73; hybrids, 67
elephants, 82; American rewilding, 150; mammoth cloning and, 134, 141, 212n3; poaching, 10–11, 93, 106–7
elk, 82
Ellis, Richard, 112
emissions, 36, 60–62; from biofuels, 69, 193n35; carbon trading, 124–25; from coal, 61, 73–74, 152, 153, 188n8, 194n47; from deforestation and other land conversions, 123; leading emissions-producing countries, 162; ocean acidification and, 46; from oil, 60–61, 188n8, 188–89n9; from Permian volcanism, 43, 184–85n15; projections and warming impacts, 36, 61–62, 185n16; transportation-related, 191n28. *See also* vehicle emissions
emissions reduction, 62–78; business initiatives, 126; carbon capture and storage for, 63, 64, 74–75; carbon-wedge scenarios, 62–65, 66; economic incentives for, 77, 78; efficiency improvements for, 63, 64, 66–69; feasibility of, 61–65; moving to carbon-neutral technologies, 62, 63, 64–66, 75–78, 168, 172; need for aggressive action, 60–62, 65, 69; quantifying economic value of, 123; vehicle emissions, 64, 66–73
endangered and threatened species, 10–12; CITES trade bans, 15, 171, 180n15; conservation success stories, 14–16, 114; current numbers, 12–13, 29; endangered status definitions and categories, 9, 11–12; IUCN Red List, 9, 180n13; "threatened with extinction" status, 8–9, 12, 13–14; "vulnerable to extinction" status, 9, 10–11, 16. *See also* wildlife *entries*; *specific animals*

Endangered Species Act (U.S.), 9
energy: energy flow and distribution among species, 53–60, 80–81; energy losses along the food chain, 55–56, 91. *See also* energy production and use; power
Energy Independence and Security Act, 67
energy production and use, xi, 60–78, 168; biofuels production, 69, 70–71, 92, 193n35, 200–201n35; energy distribution among species, 53–60, 80–81; human energy dependence and its impacts, 36, 58–60, 153–55; improving efficiency, 63, 64, 66–69, 73, 168; moving to carbon-neutral technologies, 62, 63, 64–66, 75–78, 168, 172; primary producers/net primary productivity, 52–54, 56, 57, 58, 80; reducing your individual footprint, 170–71, 175; scaling up emerging technologies, 75–77. *See also* emissions; power plants
energy solutions: need for aggressive action, 60–62, 65, 69. *See also* emissions reduction; energy production and use
environmental awareness, 126, 131, 146. *See also* consumer awareness
Erwin, Doug, 40
escolar, 104
ethanol, 69
Eubalaena (right whales), 94
Eupherusa cyanophrys (blue-capped hummingbird), 11
Europe: mileage standards, 67, 191n27; vehicle ownership, 192n32
European starling, 141, 214n17
extinction(s), ix–x, 4–9; Big Five, 13, 29–33, 48; since the Big Five, 12–13, 29–31, 32–33, 182n15, 182–83n17; of species vs. populations or subspecies, 6–7; stages in the extinction process, 7–8. *See also* K-Pg event; P-T event; Sixth Mass Extinction; *individual animal types and species*
extinction rates and magnitudes, 4, 5–6, 132; current crisis, 4–5, 13, 29–33, 48–49; hopeful signs, 13–16, 32–33; K-Pg event, 13, 19, 31; new tools to calculate, 30; P-T event, 13, 31, 37, 40; recent, 12–13, 29–32, 182n15
extinction risk, 132; habitat destruction/fragmentation and, 82–83, 88–89, 197n8; IUCN risk categories, 8–12; "prevailing circumstances" standard, 13–14; quantifying, 9, 11, 12. *See also* *individual animal types and species*
"Extraterrestrial Cause for the Cretaceous-Tertiary Extinction" (Alvarez, Alvarez, Asaro and Michel), 22–24
ExxonMobil, 71, 162, 169

failed and fragile states, 157
famine. *See* hunger
farming. *See* agriculture
Felis sylvestris (wildcats), 6, 7
fer-de-lance viper, 120–22
ferns, 39
Ferreira, Sergio, 121
fertilizer use, 89, 90; increased yields and, 165, 166; negative impacts, 48, 91, 94, 167
financial systems, 158–59, 169. *See also* economic *entries*; money
fish and fisheries, 94–105; aquaculture, 98–101, 103; commercial fishing as a way of life, 102–3; commercial fishing methods, 94, 103; consumer awareness and choices, 100, 101, 103–5, 170; economics of commercial fishing, 96, 105, 120; habitat destruction, 94; implementing catch limits, 101–3; ocean acidification and, 47; over-fishing, 94–98; Permian-era fish, 46; pollution and, 94; predator-prey

relationships, 197n7, 204n62; sport fishing, 79–80, 102; success stories, 102, 103. *See also specific fish*
flooding, 127, 153, 154
flowering plants, 39
food, xi, 29, 49, 79–105; current food-system weaknesses, 155–56; distribution among species, 80–81; energy loss along the food chain, 55–56, 91; humans as adaptable predators, 95–96; hunting of wild meat species, 93–94; predator-prey relationships, 82, 95–96, 197n7, 202–3n51, 204n62. *See also* agriculture; fish and fisheries; livestock
Food and Drug Administration, 104
food chain, energy loss along, 55–56, 91
food distribution, 155
food insecurity, 85, 93, 155–56, 164–65
food labeling, 104–5
food security solutions, 88–94, 164–68, 170; aquaculture, 98–101, 103; Green Revolution success story, 89, 164–68; increasing agricultural yields and efficiency, 88–91, 164–68; sustainable meat-eating, 89, 91–94, 170; sustainable wild fisheries, 101–5
food waste, 89, 92, 155, 170
foraminifera, 21
Ford Foundation, 167
forest elephant, 93
forests: carbon sequestration by, 123; climate change impacts, 78, 144; rainforests, 88–89, 122–24, 170, 198–99n22. *See also* deforestation
fossil fuels, 36, 60–78; agricultural inputs, 193n35; climate disruption and, 60, 61–62, 188–89n9, 190n14; current/projected consumption and emissions, 60–61, 188n8, 188–89n9, 190n12, 191n28; human dependence on, 58–60; improving efficiency, 63, 64, 66–69; moving to carbon-neutral technologies, 62, 63, 64–66, 75–78, 168, 172; sources of, 54, 55, 59; transportation-related, 66, 67. *See also* emissions; energy; power plants; vehicle *entries*; *specific fuel types*
fossils: dating methods, 20, 21–22, 31, 181n4; Montana fossil beds, 24–25, 44; South Dakota fossil beds, 17–19, 20, 181n2; techniques for identifying past carbon levels and temperatures, 40–41, 44–45, 184n8
foxes, 82
fracking, 74, 194n49
frog, growling grass, 11
fuel efficiency: of power plants, 63, 64, 66; of vehicles, 64, 66–69, 73

Galápagos Islands, 2, 3–4, 7–8, 14–16, 108
Galápagos tortoise, 1–4, 8, 10, 14–16; Lonesome George, *xiv*, 1, 2–3, 7–8, 14, 16
gardens, 92; Earth as a managed garden, 146, 150
gas mileage. *See* vehicle *entries*
gene flow, 6, 82–83
genetic engineering. *See* biotechnology; de-extinction; molecular biology
geology: geological dating methods, 20, 21–22, 25, 28, 31, 181n4; Law of Uniformitarianism, 23, 181–82n7; measuring past carbon concentrations, 40–41, 184n8
geothermal energy, 75, 76
giant tortoises. *See* Galápagos tortoise
gingko, 39
Gini coefficient, 158
Glacier National Park, 145
global warming. *See* climate change; warming
Glossopteris, 39
Goldilocks strategy, 91
Gore, Al, 162
gorgonopsids, 38–39

gorillas, 11–12, 93, 147
government: as part of the solution, 171, 172–73. *See also* economic policy; public policy initiatives; regulation
Grand Canyon, 23
grasslands biodiversity, 81, 84–88. *See also* agriculture; pasturelands
grazing, 85. *See also* livestock; pasturelands
Great Dying. *See* P-T event
Greater Yellowstone Ecosystem, 82
Great Recession, 159
greenhouse gas emissions. *See* CO_2 levels; emissions; energy
Greenpeace, 126
Green Revolution, 89, 164–68
grizzly bear, 145, 148
growling grass frog, 11

habitat conservation, 169; new perspectives on, 132–33, 146–48, 149–51; "protect and preserve" ethic, 143–46, 149
habitat destruction, 88; agriculture's impacts, 81–88, 112–13; de-extinction efforts and, 140–41; fragmentation, 82–84, 111, 145, 197n8; ocean floor, 94. *See also* deforestation; *individual animal types and species*
Hadly, Liz, 57, 86, 87
Hawaii: Kamehameha Schools land use plan, 130–31
Heatstroke: Nature in an Age of Global Warming (Barnosky), 190n18
heat waves, 153, 154
Hell Creek Formation, 24–25
Hildebrand, Alan, 25, 182n10
Hippocampus kelloggi (seahorse), 116
HMS *Beagle,* 3–5
Holser, William, 42
homing pigeons, 140
Horner, Jack, 21, 44–45
human activities, x, 28–29, 34–37, 146, 151; climate change and, 34–35, 46–47; impacts on human quality of life, 152–59; as perfect storm, 36–37, 49; "prevailing circumstances" standard, 13–14; quantifying current impact of, 29–33; recognizing as part of nature, 144–45. *See also* emissions; energy; food; money; pollution; solutions; *specific activities*
human health: air pollution impacts, 152, 153; extreme weather impacts, 153, 154
human population levels, x, 35–36, 49, 152, 175; emissions levels and, 61, 63, 190n12; energy availability and, 56–60, 80; food security and, 88, 155, 164–65; land use and, 36, 80–81, 146; per capita agricultural land requirements, 88–89
hummingbird, blue-capped, 11
hunger, 85, 155, 156, 164. *See also* food insecurity
hunting, 80, 93–94; Galápagos tortoise, 1–2; passenger pigeon, 137; trophy hunting, 110, 112; whaling, 1–2, 94. *See also* wildlife poaching
Hurricane Sandy, 34–35, 154, 163
Hutton, James, 23
hybrid vehicles, 67
hydroelectric power, 75, 76
hydrogen fuel cell-powered vehicles, 64, 72, 77
hydrogen production, 64, 72

illicit wildlife trade. *See* wildlife poaching and trade
India: Coca-Cola groundwater depletion ruling, 126; 1960s food crisis and Borlaug's work, 164–65, 166–67
Indian rhinoceros, 115
Indo-Chinese tiger, 110
Indonesia, deforestation in, 83–84, 170
Industrial Revolution, 58
infrastructure improvements, 160, 220n38. *See also* energy production; power plants

inland silverside, 47
insects, 26, 39
Integrated Valuation of Environmental Services and Tradeoffs (InVEST), 130–31
Intergovernmental Panel on Climate Change. *See* IPCC
international agreements and cooperation, 160, 163, 172; CITES, 15, 171, 180n15
International Rice Research Institute, 167
International Union for the Conservation of Nature. *See* IUCN
invasive species. *See* non-native species
InVEST (Integrated Valuation of Environmental Services and Tradeoffs), 130–31
IPCC (Intergovernmental Panel on Climate Change): AR_4 report (2007), 185n16, 190n14; AR_5 report (2013/2014), 185n16, 190n14; emissions scenarios and warming projections, 185n16, 186n24, 190n14
iridium, 22, 25
IRRI (International Rice Research Institute), 167
irrigation, 90–91
isotope excursions, as record of past carbon concentrations, 41–43
IUCN (International Union for the Conservation of Nature), 8–13, 32, 98, 103; CITES, 15, 171, 180n15
ivory poaching, 10–11, 106–7

Jablonski, Nina, 180–81n1
Jacobson, Mark, 75–76
Japanese mileage standards, 67, 191n27
Javan rhinoceros, 115
Javan tiger, 110
Johnson, Lyndon, 162, 163
Joshua Tree National Park, 146

Kamehameha Schools land use plan, 130–31
kelp forests, 197n7
Kenya: black rhino in, 112–13; ecotourism, 107–8; Maasai culture, 150–51; wildlife poaching, 106–7, 114
K-Pg event, 19–28; asteroid impact explanation for, 21–28; Chicxulub crater discovery, 25–26; climate change and, 28, 48; debates over, 22–28; disparate impacts of, 25, 26–28; energy release, 27; extinction magnitude/rates, 13, 19, 31
K-T event, 181n5. *See also* K-Pg event

lakes, 144, 197n7
land use, 36, 80–81, 146; for agriculture, 80–81, 196n3, 200–201n35; for livestock production, 80–81, 91, 200–201n35; planning tool to integrate value of ecosystem services, 130–31. *See also* agriculture
Latin American water funds, 127–29
Law of Superposition, 181n3
Law of Uniformitarianism, 23, 181–82n7
Leopold, A. Starker, 143
Leopold, Aldo, 216n33
Leopold Report, 143
Lepidocybium flavobrunneum (snake mackerel, escolar), 104
"Lethally Hot Temperatures during the Early Triassic Greenhouse" (Sun et. al.), 46
Lewis and Clark expedition, 81
Linnaean taxonomic system, 5–6, 7
lions, 82, 148, 150
lithium, 77
Litoria raniformis (growling grass frog), 11
livestock: biomass and energy requirements, 57, 58; cheatgrass and, 86–87; current land use for livestock production, 80–81, 91, 200–201n35; grazing impacts, 85; inefficiencies of livestock production, 91–92, 200–201n35;

livestock *(continued)*
wildlife and, 81–82, 147. *See also* pasturelands
livestock predators, 81–82, 147
lobster, 103
Lonesome George, *xiv*, 1, 2–3, 7–8, 14, 16
Long Now Foundation, 135
longtail tuna, 98
Loxodonta africana, 10–11, 93. *See also* elephants
Lyell, Charles, 23
lynx, Canadian, 96, 202–3n51

Maasai, 150–51
mackerel, 102
Madagascar, habitat loss in, 12
Magaritz, Mordeckai, 41–42
Maiasaura, 21
Maine lobster, 103
maize, 156
Majumdar, Arun, 65–66, 191n21
Malayan tapir, 83–84
Malayan tiger, 110
mammals: current and past extinction rates compared, 30, 32–33, 182–83n17; current extinction risk, 4, 145; K-Pg extinction survival, 26; Permian ancestors of, 38–39; Pleistocene megafauna, 93, 96, 141; small-mammal diversity and cheatgrass, 87–88; terrestrial, global carrying capacity, 55–58. *See also individual types and species*
mammoth cloning, 133–35, 141, 212n3
Manis (pangolins), 116
marine life: dead zones, 47–48, 71, 94; impacts of ocean acidification, 46–47; K-Pg asteroid impact and, 27–28; Permian-era, 39–40, 46; as primary producers, 54, 55. *See also* fish and fisheries
marine reserves, 149, 150
marine sediments, 20, 55, 184–85n15
Marris, Emma, 146
Martin, Jim, 17–18, 180–81n1
Martin Paleontology Research Laboratory, 17–18, 180–81n1
mass extinctions. *See* extinction *entries*; *specific events*
maximum sustained yield (fisheries), 102
McKenna, Malcolm C., 26–27
McKenna, Priscilla, 26–27
meat: eating less, 89, 91–92, 170; overhunting of wild meat species, 93–94. *See also* livestock; pasturelands
medicine: pharmaceuticals development, 120–22; wildlife poaching for medicinal uses, 109–17
megafauna: Earth's megafaunal carrying capacity, 55–58; mammoth cloning, 133–35, 141, 212n3; Pleistocene megafauna, 93, 96, 141; recent diversity declines and extinctions, 57–58. *See also individual types and species*
Melton, Bill, 44–45
Merchants of Doubt (Oreskes and Conway), 161
metapopulations, 83
methane, 36, 184–85n15
Mexico: Borlaug's work in, 165–66; Chicxulub Crater discovery, 25–26
Michel, Helen V., 22. *See also* asteroid-impact extinction theories
Michener, Charles, 129–30
mileage and mileage standards. *See* vehicle *entries*
minimum viable population size, 83, 197n8
mining: coal mining, 50–52, 58–59, 78; minerals for emerging energy technologies, 77
"Miracle Rice," 167
molecular biology, 132–33, 142. *See also* de-extinction
money, xi, 29, 49, 106–31, 156–59; climate-change denial and, 161–62; costs of de-extinction efforts, 140,

213–14n15; costs of extreme weather, 153; economics of commercial fishing, 96, 105; poverty and economic inequality, 156–59; short-term vs. long-term value, xi, 108–9, 120, 122–24; water funds, 127–29. *See also* consumer awareness; economic systems; natural capital; wildlife poaching and trade
Montana: fossil beds, 24–25, 44; pasturelands, 84–85
Monterey Bay Aquarium Seafood Watch, 103, 170
mosasaurs, 17, 18, 20–21, 27–28
Moschus (musk deer), 116
Mozambique, white rhino in, 114
musk deer, 116
mustangs, 147

National Geographic, 135
national parks, 145, 149; African, poaching in, 10, 11, 107, 114; Galápagos, 14; the Leopold report, 143. *See also specific parks*
natural capital, 49, 122; short-term vs. long-term exploitation of, xi, 108–9, 120, 122–24; water as, 126, 127–29; wildlife as, 106, 107–9. *See also* ecosystem services
Natural Capital Project, 130
natural gas, 59, 61, 188n8; fracking, 74, 194n49; natural gas-fired power plants, 64, 74. *See also* fossil fuels
Naylor, Rosamond, 99–100
negative $\delta^{13}C_{carb}$ excursions, 41, 42–43, 184n8
Nelson, Steve, 180–81n1
net primary productivity (NPP), 54, 57, 80
New York City: extreme weather and, 34–35, 154, 163, 220n43; water supply, 127
nitrogen runoff and accumulation, 48, 71, 91, 94, 144
nomenclature, 5–6, 7
non-native species: aquaculture and, 99; cheatgrass invasiveness, 85–88
northern white rhino, 113–14
Nowak, Ronald, 112
NPP (net primary productivity), 54, 57, 80
nuclear power, 64, 75
nutrient runoff, 47–48, 71, 91, 94, 144

Obama, Barack, 163, 172
oceans: acidification, 37, 46–47, 48; anoxia and dead zones, 47–48, 55, 71, 94, 99; energy production in, 54, 55; marine reserves, 149, 150; ocean-based energy technologies, 75, 76, 195n58; ocean temperatures, 35, 41, 44, 45–46; Permian-era oceans, 41, 44, 45–47; sea level rises, 37, 154; warming and storm intensities, 35. *See also* fish and fisheries; marine *entries*
OECD countries, economic inequality in, 158
oil, 55, 59, 60–61; current consumption and emissions from, 60–61, 188n8, 188–89n9, 190n12; new extraction methods, 74, 194n49, 195n60; oil supplies, 77, 195n60. *See also* fossil fuels
oil industry, 162, 169
Oncorhynchus tshawytscha (Chinook salmon), 100
"Opportunities and Challenges for a Sustainable Energy Future" (Chu and Majumdar), 65–66, 191n21
Ordovician: end-Ordovician extinction, 31, 48
Oregon climate-change initiatives, 172
Oreskes, Naomi, 161
Organisation for Economic Co-operation and Development countries, economic inequality in, 158
Oryx, 116
overgrazing, 85

oxygen: anoxia, 47–48, 55, 71, 94; past temperature estimation method, 44–45
oysters, 47

Pacala, Stephen, 62–65, 66
Pacific bluefin tuna, 96, 97, 103, 105
Pakistan, Borlaug's work in, 166
Paleogene, 21. *See also* K-Pg event
paleontology: dating methods, 20–22, 25, 28, 31, 181n4; debates over asteroid-impact theory for K-Pg event, 22–28
palm oil, 84, 170
pangolins, 116
Panthera (tigers), 7, 109–11, 116, 120, 148
Pan troglodytes (chimpanzee), 93
passenger pigeon, 136–41, 213–14n15
past extinctions. *See* extinction(s); *specific events*
pasturelands, 84–88, 91–92. *See also* livestock
Patagioenas fasciata (band-tailed pigeon), 138, 139, 213–14n15
Payne, Jon, 42
peak oil debate, 77, 195n60
Pentaceratops, 21
"perfect storm," 35, 183n1; human activities as, 36–37, 49
Permian fauna, 38–39
Permian-Triassic extinction event. *See* P-T event (Great Dying)
pesticide use, 89, 90, 166, 167
pharmaceuticals development, 120–22
Phelan, Ryan, 135; Revive and Restore project, 135–41
Philippines: International Rice Research Institute, 167
photosynthesis, 52–54, 58, 59
Pierre Shale, 18–19, 20, 181n2
pigeons: passenger pigeon de-extinction efforts, 136–41, 213–14n15
Pimm, Stuart, 142
plankton, 27–28, 197n7
plants: as energy producers, 52–54, 58, 59; K-Pg extinction survival, 26; Permian-era, 39. *See also* forests
Pleistocene megafauna, 93, 96, 141; mammoth cloning, 133–35, 141, 212n3
Pleistocene rewilding, 150
plug-in electric vehicles, 72–73
poaching. *See* wildlife poaching
politicians, as part of the solution, 171, 172–73
politics, 171; climate-change denial and, 162, 164; political instability, 157
pollution: from aquaculture, 99, 101; dead zones, 47–48, 71, 94, 99; health impacts of air pollution, 153; nutrient runoff and accumulation, 47–48, 71, 91, 94, 144; from pesticide and fertilizer use, 48, 91, 94, 167. *See also* emissions *entries*
population, human. *See* human population levels
populations: habitat fragmentation impacts, 82–84, 111, 197n8; metapopulations, 83; minimum viable population size, 82–83, 197n8; vs. subspecies, 6–7
porpoises, 202n46
poverty, 156–57, 218–19n26
power (energy), xi, 29, 49, 50–78, 152–55. *See also* emissions; energy; fossil fuels
power (to act), xi, 14, 16; taking action, 169–76. *See also* consumer awareness and choice; solutions
power plants: associated air pollution, 152, 153; biofuel-powered, 74–75; coal-fired, 64, 73–75, 153; natural-gas facilities, 64, 74; reducing emissions from, 63, 64, 66, 74–75. *See also* energy production
predator-prey relationships, 82, 95–96, 197n7, 202–3n51, 204n62
"preserve and protect" ethic, 143–46, 149
primary producers and productivity, 52–54, 56, 57, 58, 80

The Principles of Geology (Lyell), 23
pterosaur, 19
P-T event (Great Dying), 31, 37–47, 183n3; atmospheric CO_2 increase as trigger, 38, 40, 43; extinction magnitude, 13, 31, 37, 40; ocean conditions, 46–47; Permian fauna and flora, 38–40; related warming, 43–46
public policy initiatives, 172–73; carbon trading, 125–26; recent climate change initiatives, 163, 169, 172; reforestation initiatives, 127; water funds, 127–29. *See also* economic policy; regulation

Quito Water Fund, 128–29

radiometric dating methods, 20, 21–22, 25, 181n4
rainforests, 88–89, 120–24, 170, 198–99n22. *See also* deforestation
rare-earth metals, 77
REDD program, 125
Red List of Threatened Species, 9, 180n13. *See also* IUCN
Reducing Emissions from Deforestation and Forest Degradation (REDD program), 125
regulation: aquaculture industry, 100, 101; commercial fishing, 101–3; fuel efficiency standards, 67; seafood labeling/identification, 104–5
religion, climate-change denial and, 162
Renne, Paul, 28
reptiles: Permian ancestors of, 39, 46, 183n5. *See also individual types and species*
restaurants, fish misidentification in, 104–5
resuscitation, 132–51; classic conservation biology, 143–46; de-extinction efforts, 133–42, 149, 213–14n15; new directions in conservation biology, 132–33, 146–48, 149–51
Revive and Restore, 135–36; passenger pigeon project, 136–41, 213–14n15
rewilding, 150
rhinoceros, 109, 111–15, 118–19, 147, 149
rice, 156, 167
Ricketts, Taylor, 129–30
right whales, 94
Rockefeller Foundation, 164, 167
rock pigeon, 139–40
rocks: dating and matching of, 20, 21–22, 25, 28, 31, 181n4; techniques for measuring past carbon levels and temperatures, 40–41, 44–45, 184n8
rodents, cheatgrass and, 87–88

Sahel, 85
salmon and salmon aquaculture, 98–100, 103, 204n62
Salmo salar (Atlantic salmon), 98–100, 103
Sapphire Energy, 71–72
scientific nomenclature, 5–6, 7
scientists: call to action, 174
Scott, Harold, 45
Scottish wildcat, 6, 7
seafood labeling and identification, 104–5
Seafood Watch, 103, 170
seahorses, 116
sea level rises, 37, 154
seals, 116
sea otter, 197n7
sea urchins, 197n7
sewage. *See* wastewater
Sharon Springs Formation, 181n2
Siberian tiger, 110
Siberian Traps, 43, 184n14
Siemens Healthcare, 126
Sixth Mass Extinction, x–xi, xii, 32, 48–49, 62; compared to Big Five Mass Extinctions, 29–33; magnitude of, 4–5, 13, 29–33. *See also* solutions
snake mackerel, 104

snow leopard, 116, 149
snowshoe hare, 96, 202–3n51
social media campaigns, 116, 117
Socolow, Robert, 62–65, 66
solar energy, 52, 172; amount available, 53, 187n1; for hydrogen production, 72; increasing our use of, 64, 75, 76; photosynthesis, 52–54, 58, 59; solar lights, 175, 223n67
solutions, x–xi, xii, 14, 16, 159–76; acknowledging the problems, 14, 49, 159, 161–64, 169; economic solutions, 168–69; ecosystem services valuation/integration, 124–31, 146, 149, 168; fighting wildlife poaching, 93, 116–20, 131; food supply and security, 88–94, 164–68, 170; key elements of, 159–61, 167–68; new directions in conservation biology, 132–33, 146–48, 149–51; past success stories, 76, 89, 160, 164–68, 220n38; reducing emissions and fossil fuels dependence, 60–78, 169; taking action, 169–76. *See also* consumer awareness and choice; emissions reduction; food security solutions
"Solutions for a Cultivated Planet," 89–90
Soulé, Michael, 216n33
South Africa, rhinoceros in, 114
South China tiger, 110
South Dakota fossil beds, 17–19, 20, 181n2
South Dakota School of Mines, 17–18, 180–81n1
southern bluefin tuna, 96–97, 103
southern white rhino, 114, 147
soybeans, 122–24, 126, 156
species: defined, 5; number of, 9
sport fishing, 102
starlings, 141, 214n17
Steadman, David, 182–83n17
Stegner, Wallace, 216n33
storm intensities, 34–35, 37, 153–54, 163. *See also* climate change
students: call to action, 173–74
Sturnus vulgaris (European starling), 141, 214n17
subspecies, 5, 6–8
success stories: agricultural improvements, 89, 164–68; aquaculture, 100–101; fisheries sustainability, 102, 103; technological/infrastructure breakthroughs, 76, 160, 220n38; wildlife conservation, 14–16, 114
Sumatran rhinoceros, 115
Sumatran tiger, 110
Sun Yadong, 45–46
Superstorm Sandy, 34–35, 154, 163
sushi restaurants, fish misidentification in, 104–5
swordfish, 102
synapsids, 38–39
syngas, 64
synthetic biology, 70. *See also* biotechnology; de-extinction; molecular biology
Synthetic Genomics, 71

Talbot, Bob, 180–81n1
Tanzania, elephant poaching in, 10
Tapirus indicus (Malayan tapir), 83–84
tar-sands oil extraction, 195n60
Task Force on Climate Preparedness and Resilience, 172
technology and technological innovation, 160–61, 169, 175; agricultural technologies, 90–91, 166–67; anti-poaching technologies, 118–19; communications technologies, 49, 116, 117, 160–61; DNA identification methods, 104–5, 119; feasibility of innovation, 76, 160, 220n38; scaling up new technologies, 75–77, 167. *See also* biotechnology; de-extinction; molecular biology
TED de-extinction conference, 135

temperatures: after K-Pg asteroid impact, 27–28; current ocean temperatures, 44, 45; past, techniques for estimating, 41, 44–45; twentieth-century Earth surface temperatures, 44, 186n18. *See also* warming
Tercek, Mark, 126
Tertiary, 181n5
threatened species: "threatened with extinction" defined by IUCN, 8–9, 12, 13–14. *See also* endangered and threatened species
Thunnus (tuna), 11, 96–98, 102, 203n53
tidal energy, 75, 76
tigers, 7, 109–11, 116, 120, 148
tilapia, 98, 100–101
tobacco industry, 161
tortoises. *See* Galápagos tortoise
traditional medicine: wildlife poaching for medicinal uses, 109–17
transportation-related fossil fuels use, 66, 67. *See also* vehicle *entries*
Triassic: carbon excursions during, 42; end-Triassic extinction, 31, 48. *See also* P-T event
trilobites, 40
tsetse fly eradication, 112–13
tuna, 11, 96–98, 103, 104, 105, 203n53
turtles, 26, 116

U.S.: agricultural efficiency improvements, 166; climate-change denial and acknowledgment, 162–63, 164; economic inequality in, 158; as emissions leaader, 162; government climate-change initiatives, 163, 169; poverty and hunger in, 156, 218–19n26
U.S. Food and Drug Administration, 104
U.S. legislation: Endangered Species Act, 9; Energy Independence and Security Act, 67
U.S. military biofuels development, 69–70

UNESCO, 15
Ursus thibetanus (Asiatic black bear), 116–17, 120

Vane, John, 121
vehicle emissions, 64, 66–73; alternative engine technologies, 64, 65, 72–73; biofuels for, 64, 69–72, 193n35; current emissions figures, 68, 191n28, 192nn29, 30, 31; fuel efficiency standards, 67, 191n27; improving fuel efficiency, 64, 66–69
vehicle ownership and use, 64, 68, 192n32
Venrock, 71
Vietnam, illicit wildlife trade and, 113, 115
Vision 2050, 168–69, 222n60
volcanic activity, CO_2 emissions from, 43, 184–85n15
"vulnerable to extinction" status, 9, 10–11; Galápagos tortoise's move to, 16
Vultur gryphus (Andean condor), 128, 211n61

warming, 35, 36, 183n2; anoxic waters and, 48; Big Five extinctions and, 43–46, 48; current, impacts on nonhuman species, 63, 190n18; current and future, impacts on humans, 154–55, 156; current fossil fuel use and, 61–62, 188–89n9, 190n14; future scenarios and projections, 43, 63, 185n16, 186n24, 190n14; land use conversions and, 123; mammoth extinction and, 141; Permian-era, 43–46; "safe" warming target, 63
Washington state climate-change initiatives, 172
wastewater, 47–48, 71
water: agricultural use, 90–91; as natural capital, 126, 127–29; potential shortages, 154–55; water funds, 127–29. *See also* oceans

water quality: aquaculture and, 99, 101; nutrient runoff and ocean dead zones, 47–48, 71, 94, 99; in remote ecosystems, 144
wave energy, 75, 76, 195n58
western gorilla, 11
Western Interior Seaway, 18
whales and whaling, 1–2, 94, 202n46
wheat, 156, 165–67
white rhinoceros, 112, 113–14, 147
wildcats, 6, 7
wilderness areas/wild places, 144–51, 216n33; future of protected areas, 149–50; marine reserves, 149, 150; new perspectives on, 146–48, 150; "protect and preserve" ethic, 143–46, 149
wildfires, 28, 86, 153
wildlife conservation, 15, 175; captive breeding programs, 15; ecosystems approaches, 142; future needs and solutions, 147–48, 149–50, 151; in the Galápagos, 14–16; Galápagos tortoise, 14–16; Indian rhinoceros, 115; marine reserves, 149, 150; motives for, 148; predator control, 15–16; rewilding, 150; social media campaigns, 116–17; success stories, 14–16, 114; water funds and, 128. *See also* conservation biology; de-extinction
wildlife farming, 93, 119–20, 149
wildlife poaching and trade, 10–11, 106–20, 145–46; bush-meat trade, 93–94; CITES, 15, 171, 180n15; elephants, 10–11, 106–7; fighting, 93, 116–20, 131; Galápagos tortoise, 14, 15; habitat fragmentation and, 111; impacts of, 93–94, 106–16; rhinoceros, 109, 111–15; tigers, 109–11
wild pigeon (passenger pigeon), 136–41, 213–14n15
Williams, Terry Tempest, 216n33
wind energy, 64, 72, 75, 76, 172, 195n58
wolves, 82, 147, 149
woolly mammoth cloning, 133–35, 141, 212n3
World Business Council for Sustainable Development, 168–69, 222n60
World Conservation Union, 15
World War II, 76, 160, 164, 220n38
World Wildlife Fund, 126, 129

Xi Jinping, 163

Yamani, Ahmed, 77
yellowfin tuna, 97–98, 105
Yellowstone ecosystem, 82
Yellowstone National Park, 108, 144, 145
Yemen, as market for rhinoceros horn, 113
yields and yield gap (agriculture), 89, 90, 155, 156, 164–68
Yochelson, Ellis, 45
Yosemite National Park, 108
youth: call to action, 173–74

Zangerl, Rainer, 45
Zimbabwe, elephant deaths in, 11
zoos, 9, 140, 141, 142, 148